Psychobiological Footprints through Human Development

IIII I IIIIIIIIIIIIIIIIIIIIIII IIII IIII
I0028273

Embark on an illuminating voyage through the biological foundations of human nature and development with *Psychobiological Footprints through Human Development*. This unique volume unveils the intricate dance between genetics, neuroscience, and environment, offering a holistic understanding of how we become who we are.

This comprehensive book examines the psychobiological, neuroendocrine, and epigenetic mechanisms that regulate developmental processes in typical development and under conditions of developmental risk. Moving within a dynamic systems epistemic framework and capitalizing from the heritage of the infant research field, it provides a solid framework for comprehending the interplay of nature and nurture. With a captivating blend of theoretical principles, processes, and contextual applications, this book transcends academic boundaries to empower anyone interested in the intricacies of human development.

Psychobiological Footprints through Human Development is a guide to discovering how our life experiences contribute to making us who we are and therefore it is invaluable to graduate students in the fields of developmental psychology, neuroscience, genetics, and related disciplines. Delving into the biological roots of behaviour, cognition, and emotion, it will also equip practitioners, researchers, and educators with invaluable insights to enrich their practice.

Livio Provenzi, psychologist and psychoanalyst, holds a PhD in Psychology from the Catholic University of Milan and a specialization in Psychotherapy from the Italian Society of Relationship Psychoanalysis. Author of numerous publications in international scientific journals, Livio Provenzi is Associate Professor in the Department of Brain and Behavioural Sciences of the University of Pavia, where he teaches Developmental Psychobiology and Developmental Psychopathology. He also directs the Developmental Psychobiology Lab of the Mondino Foundation, Pavia (Italy).

Psychobiological Footprints through Human Development

How Our Experiences Shape Who We Are

Livio Provenzi

Routledge
Taylor & Francis Group

LONDON AND NEW YORK

Cover image: Hulinska Yevheniia via Getty Images

First published 2025
by Routledge
4 Park Square, Milton Park, Abingdon, Oxon OX14 4RN

and by Routledge
605 Third Avenue, New York, NY 10158

Routledge is an imprint of the Taylor & Francis Group, an informa business

Translated from the Italian *Psicobiologia dello sviluppo. Principi, processi e contesti* © 2021 by Carocci editore, Roma

British Library Cataloguing-in-Publication Data
A catalogue record for this book is available from the British Library

ISBN: 9781032766188 (hbk)
ISBN: 9781032766157 (pbk)
ISBN: 9781003479314 (ebk)

DOI: 10.4324/9781003479314

Typeset in Times New Roman
by KnowledgeWorks Global Ltd.

Contents

Prologue   *ix*

Where do we come from? ix
What will you find in this book? xi

SECTION I
Theory   1

1 Order and disorder: Non-linear dynamics of development   3

When we started observing 3
The system 5
Organization and specificity: The centrality of time 8
Inevitable ruptures 11
Matching, mismatching, and other dance steps 13
Meaning making 15

2 Born to be wired   17

In Taung (South Africa) 17
In utero 19
Just like me 20
Giant cells 22
The skin as a place to meet 24
Conclusions 25

3 The assembly grammar   26

To select and to instruct 27
Not in a vacuum 29
Frog thighs and hearts 31
Chemical brothers 34

The assembly 35
Learn like a slug 36
Beyond learning: Recollection 37
An integrated systemic view of developmental psychobiology 40

SECTION II
Processes   43

4 Social by evolution   45

The Da Vinci code 45
The vagal paradox 47
Of reptiles and mammals 48
Polyvagal 49
Distinguishing foes from enemies 50
Critiques 51
In unison 52

5 The culprit of stress   55

Let's stress it out 55
A neuroendocrine cascade 57
Everyday resilience 58
Pump up the volume or shut the system down? 60
What makes a good rat mama 62
Too much and too little 63
Paying the consequences 64
Parenting as a protective buffer 67
Not only stress: The role of HPA in memory and learning 68

6 Genes and what you do   71

Genetics and psychology, a difficult relationship? 72
From quantitative genetics to molecular genetics 73
Vulnerable to stress 75
Evolutionary reasons 76
Psychobiological reasons 78
Better than vulnerable: Malleable 79
Conditional adjustments 80
Temperament and sensitivity 82
Genetic variations 82
Positive and negative: Does it make sense? 86

7 Footprints in the epigenome 88

Paris and London 88
Looking at another landscape 89
When environments shape genes 90
Stories of rodents 92
Hallelujah 94
Behavioral epigenetics 96
The seductive charm of a promise 98

8 Other frontiers 100

Stress that can inflame 100
A second brain 102
Powerful organelles 105
As long as life 107

SECTION III
Contexts 109

9 Waiting for 111

The placental barrier 112
Beyond the barrier 113
Caregiving programming 115
Prenatal conflicts 117

10 Born soon 120

Dark side 121
Protecting the newborn and the parents 122
Music power 124
Epigenetics and neuroplasticity 125
Behavioral epigenetics of prematurity 126
Light side 128

11 Embodied parenting 130

Psychobiology of maternal caregiving 130
Is there a paternal brain? 132
Hormonal plasticity 134
Synced dancers 135
Interbrain wires 138

12 Memories of trauma 140

Psychobiological scars 142
9/11 144
Below zero 145
The hunger winter 146
Inheriting a nightmare 147
Concluding remarks 148

**Epilogue: Framing developmental psychobiology
in complexity** 149

Complex 149
Wired 149
Quasi-adapted 150
Intentional 151
What's wrong? 152
How do we proceed from here on out? 154

**Notes on contemporary issues, on frailty and care:
Psychobiological postcards from the pandemic** 156

Acknowledgments 159

References *160*
Index *204*

Prologue

Where do we come from?

The contemporary conception of the child, widely acknowledged across various domains of clinical, developmental, and experimental psychology, stems from two major revolutions that unfolded between the twentieth and twenty-first centuries. Until the 1970s and 1980s, terms such as "interaction" were stigmatized, "intersubjectivity" was absent from discourse, "subjectivity" carried an obscure connotation, and the notion of a "theory of subjectivity" seemed implausible in many applied clinical fields of psychological and psychiatric sciences. At present, these once-neglected concepts have become commonplace in the lexicon of psychologists, whether they are involved in clinical and psychotherapeutic practices or are engaged in research within the socio-relational and socio-cognitive realms.

The first of these transformative movements can be attributed to the pioneering works of Louis Sander. In Boston, he conducted seminal research involving direct, systematic, and analytical observation of interactions between infants and their mothers. Relevant contributions also emerged from Daniel Stern, Ed Tronick, Colwyn Trevarthen, and Beatrice Beebe, whose well-known studies significantly influenced the field. These pioneering efforts, along with other noteworthy researchers mentioned in the literature, catalyzed a shift in focus from a predominantly individual-centric approach to a profound emphasis on the significance of infants' interactions within the caregiving environment, a largely relational and human context that encompasses both social and physical aspects. In essence, this first revolution, roughly spanning from the 1950s to the 1980s, led to the establishment of one of the most vibrant and enduring research strands in the field of psychology, known as the Infant Research tradition. This line of investigation remains actively pursued to present times and profoundly permeates various applicative domains of clinical and developmental knowledge within the psychological field.

A second revolution emerged in the 1990s and the first two decades of the new century, and it still can be considered an ongoing process. This transformative phase witnessed the progressive expansion and integration of knowledge that was previously confined within the established boundaries of Infant Research to include influences and interactions from external insights beyond the realm of psychology. Researchers became increasingly interested in exploring the mechanisms regulating early interactions between parents and children, extending their investigations beyond the behavioral aspects of interactive partners to encompass physiological and neurobiological processes. The objective was to elucidate individual variations and behavioral connections.

An illustrative example of this expanded perspective is the concept of the "transmission gap," particularly emphasized by Marinus van IJzendoorn (1995; de Wolfe & van IJzendoorn, 1997). As one of the leading attachment researchers, van IJzendoorn's work shed light on the discrepancy between theoretical expectations and empirical findings concerning the relationship between maternal sensitivity and attachment security. Attachment theory posits a substantial positive correlation between maternal sensitivity and the likelihood of a child developing a secure attachment style. However, empirical data only demonstrate a partial association, and meta-analytic studies reveal a modest effect size statistically, indicating a reduced strength of the association due to variations in maternal sensitivity. Consequently, it becomes challenging to support a strict causal hypothesis regarding the role of maternal sensitivity in a child's attachment. The shift in focus away from solely environmental influences toward considering individual predispositions, such as genetic or neuroendocrine factors, has gained prominence within a complexity-inspired dynamic system viewpoint of human development and parenting (see Chapter 1 of this volume). This perspective has indirectly paved the way for the embedding of the theory of differential susceptibility within the field of developmental psychology (Belsky, 2002), which represents a notable example of successful hybridization among evolutionary theories, psychological and biological approaches to understanding human development.

However, the ongoing second revolution, as previously mentioned, remains incomplete, and even the involvement of genetics has yielded only partial and unsatisfactory results. Although it has opened up new avenues and hypotheses concerning the socio-emotional and interactive aspects of child development and their caregivers, the study of individual polymorphisms failed to explain significant portions of variance in child behavior (Del Giudice, 2017). Furthermore, the joint analysis of multiple polymorphisms necessitates sample sizes rarely available in psychology. Consequently, attention shifted not so much toward structural genetic variations, but rather toward understanding how interactions with the caregiving environment could modulate DNA functioning. This revolutionary perspective, known as behavioral epigenetics, has reshaped our understanding of the interplay between nature and nurture, genetics, and culture. Furthermore, it moved the field from the quest to determine psychological traits inheritance (quantitative behavioral genetics) to the unveiling of specific molecular mechanisms that underpin the interaction of genetic predisposing factors and environmental exposures and their joint influence on developmental trajectories and individuals' phenotype (developmental behavioral epigenetics).

The scientific landscape marked by these successive revolutions—Infant Research and developmental behavioral epigenetics—outlines the contours of a scientific arena known as developmental psychobiology. This term refers to the investigation of neurobiological processes contributing to individual variations in child development within their life context. Understanding how the environment and biology can program more or less adaptive developmental trajectories in early life represents a central focus. Developmental psychobiology stands at the intersection of diverse disciplines—psychology, biology, neurophysiology, genetics, and neuroendocrinology—and benefits from a fruitful convergence of validated and innovative methodologies, generating intriguing and promising research hypotheses. This area of scientific study provides specific access points to understanding developmental processes, particularly during early childhood. Humans, being immensely complex, often transcend the limits of our observational tools and theoretical frameworks. By paraphrasing the words of my mentor Rosario Montirosso, "humans are far more complex than theoretical models." As we acknowledge such incontrovertible assumption, developmental psychobiology acts as a catalyst for our attempts to (at least partially) comprehend human development, with profound implications for scientific research, clinical practice, and training.

However, the rapid accumulation of evidence in developmental psychobiology can prove challenging for those approaching it from outside the field. From an external perspective, this area of research may appear somewhat fragmented or disorderly due to its continuously evolving sub-strands, addressing a wide array of research questions. Therefore, this book seeks to serve as an introductory guide to developmental psychobiology, elucidating its fundamental principles, examining some of the most investigated biological and neurophysiological mechanisms, and exploring their application in clinical and developmental psychology contexts.

What will you find in this book?

The first section (Principles) comprises three chapters. The initial chapter introduces key concepts from Infant Research, establishing an epistemological framework that enriches the subsequent understanding of biological processes. The second chapter highlights advances that offer valuable insights into the study of intersubjectivity from a neuroscientific and neurobiological standpoint. The third chapter presents fundamental neurophysiological principles to provide a solid foundation for the ensuing content.

The second section (Processes) presents a series of biological macrosystems that have fostered intriguing and productive integrations with psychology in recent decades. Topics include parasympathetic regulation and the polyvagal theory (Chapter 4), the neuroendocrine stress regulation system, the hypothalamic-pituitary-adrenal axis (Chapter 5), genetics (Chapter 6), and behavioral epigenetics (Chapter 7). Chapter 8 summarizes innovative contributions related to the immune system, telomere biology, the microbiome, oxidative stress, and mitochondrial regulation.

The third section (Contexts) consists of non-exhaustive literature reviews to elucidate the psychobiological mechanisms and processes in action in specific developmental populations. Themes covered include prenatal psychobiology (Chapter 9), the preterm-born child (Chapter 10), trauma related to individual histories or large-scale events (Chapter 11), and embodied parenthood in men and women (Chapter 12). Each chapter emphasizes how various psychobiological processes elucidated in the book can aid in understanding the impact of environmental exposures on human development. These processes can facilitate the initiation of protective mechanisms or, conversely, lead to early risk factors for dysfunctional behavioral phenotypes.

The concluding section (Implications) highlights the idea that biological alterations in different macrosystems can be regarded as intentional biological learnings throughout an individual's life course. This perspective offers essential insight for clinical thinking and working with children, families, and adult patients. The final chapter (epilogue) briefly reflects on a recent and unprecedented healthcare emergency—the COVID-19 pandemic—suggesting that developmental psychobiology can contribute not only to understanding the effects of stress but also to initiating preventive and care interventions for the most vulnerable. The development of a vulnerability-focused culture can redefine concepts of risk and resilience from a preventive standpoint, and developmental psychobiology plays a crucial role in this regard.

As a final note, readers will most probably agree with the fact that nowadays manuals become old very fast, as the scientific progress is faster than ever before. As such, this book has been built not only to be an access manual to developmental psychobiology, but is also meant to serve as an orienting tool, guiding those who seek a scientific approach to delve into and comprehend the realm of developmental psychobiology. It also aspires to provide thought-provoking insights for students and professionals working in clinical and research settings, encouraging them to engage in this captivating area of scientific exploration. I envision this book as a guide for curious travelers, a hitchhiker's guide to developmental psychobiology. I did not want to (nor could I have) included everything in it. Instead, I chose to draft adventurous paths and navigation tips that could provide a meaningful framework, basic tools and knowledge, as well as windows of applicability, allowing travelers to pause and delve deeper into the understanding of certain aspects of the psychobiological landscape and to start from this volume to build smarter science and care for children and families.

Section I
Theory

1 Order and disorder

Non-linear dynamics of development

This initial chapter aims to furnish interpretative and epistemological tools for contextualizing the early development of the human beings and the regulatory dynamics characterizing it. Development, contrary to a linear, predetermined, or preordained process, is characterized by inherent disorder, involving a continuous negotiation among regulatory processes governing various physiological and behavioral systems. This ongoing negotiation culminates in what is known as the "phenotype." Development, thus, exists in a state of disorder and dynamic equilibrium, perpetually participatory and responsive to the influences of the child's environment—an essential environment, regardless of its specific attributes, for the very existence of the organism and the prerequisites necessary to initiate, sustain, and regulate developmental processes.

The viewpoint outlined in this chapter, which Tronick (2017) has aptly termed as "messiness," serves to establish the boundaries within which I will present developmental psychobiology. Such a perspective proves indispensable in conferring meaning to the implications that various psychobiological systems explored in Section II hold for the early development of the child. It situates the relational context, or rather, the interactive level, as the focal point for analysis, observation, and construction of meanings, essential to attain a coherent perspective of developmental psychobiology.

When we started observing

There is a specific moment in the history of human development epistemology—particularly concerning early development, let's say within the first three years of life—that marks an irrevocable turning point. This moment occurred when developmental psychology adopted a scientific approach that entailed systematic observation of early interactions between a parent and their child. Louis Sander, a giant on whose shoulders we can seat, emerged as the pioneering figure behind this observational, scientific, and systematic approach to studying early interactions. Sander, a psychoanalyst, psychiatrist, and researcher, is undoubtedly the father of Infant Research as a science and the inspiration behind a study group that played a significant role in redefining not only developmental theories but also their implications for psychotherapy and the theory of change (Boston Process of Change Study Group, 2012).

DOI: 10.4324/9781003479314-2

In truth, before Sander's contributions, the groundwork had already been laid by the now-classic studies on the effects of early affective deprivation conducted by Spitz (1945). Specifically, Spitz observed the children who were cared for in orphanages, comparing them to those cared for in a nursery attached to a women's prison. The children in the nursery received care from their mothers, while the orphanage had a caregiver-to-child ratio of about 1:7. It became evident that the children in the orphanage experienced reduced human contact—both physical and emotional—in terms of quantity and timing. In addition, the nursery's open cribs allowed children to observe their surroundings, mothers, other children, and ward activities. In contrast, the bars covering the cradles in the orphanage greatly limited environmental stimulation. The socio-sensory deprivation experienced by the orphanage children led to what Spitz termed as "hospitalization syndrome," akin to what can be referred to as anaclitic depression in adults. By the age of 8 months, children raised in orphanages exhibited greater deficits in socio-affective, motor, and cognitive development, which became more evident in the second and third years of life, contributing to a generalized delay in development. In Spitz's longitudinal study, only 2 out of 26 children walked and spoke by the second year of life.

Sander's studies certainly inherited the revolutionary scope of Spitz's work, particularly the idea that the caregiving environment plays a crucial role in influencing psychological, cognitive, emotional, and motor developmental processes. This becomes especially relevant when considering the paradox that motivated Sander to undertake the first systematic study on parent-child interaction: on the one hand, there is the uniqueness of each individual, and on the other hand, there is the potential for individual and systemic changes that enable coherence and continuity in life. In other words, how can individuals be distinct from others while remaining interconnected?

Sander initiated his research on mother-child dyads around the 1950s and 1960s in Boston aiming at observing and describing the complexity of parent-child interaction over time. Through extensive observations of 22 mother-child dyads, followed from birth until the third year of life, Sander coined the concept of *elective affinity regulation* to describe the process of mutual adaptation between parent and child. This highlights a distinctive aspect of Sander's thinking, revealing his exceptional acuity in making the complexities of human development comprehensible. The concept of adaptation is considered not passive—neither the parent nor the child merely adapts to their care environment—but rather, elective affinities involve active, ongoing adaptations, never definitively stabilized, and sensitive to the qualities of both the environment and the child. This dynamic aspect of adaptation is evident in the difference between the concepts of *adaptation* and I, where the latter conveys an active and "in action" dimension of a process of regulation that is never concluded, rather continuous in motion—a process that can be perceived as somewhat untidy.

Sander identified at least seven stages over the first three years during which parent and child regulate their "elective affinities." The first 3 months focus on stabilizing physiological rhythms (initial regulation), while the emergence of the social smile shifts the focus of mutual regulation between 4 and 6 months to reciprocal interactive exchange. Between 7 and 9 months, the child takes on an increasing communicative initiative and their first directed activities become evident. Toward the end of the first year, Sander identifies the "focus" phase, characterized by the ability to walk, enabling exploration of the world and separation from the reference figure, which sets the stage for the emergence of a primary affective relationship with the caregiver and the need for a secure base. At around 18–20 months, a subsequent phase of "self-assertion" is identified, marked by exuberant growth in the child's verbalizations, seeking greater autonomy, intentionality, and attempts at separation. Between 18 and 36 months, the "recognition" phase takes the stage, reflecting shared awareness of being in interaction, akin to a first theory of mind, and recognition of the constancy of the self across time and space, capable of resisting even moments of separation. The negotiation of these moments within the dyad is influenced by the child's endogenous initiative and the availability of the care environment. Sander describes the observable, behavioral outcomes of these regulations as the *assembly grammar* of the structure of events, representing how mother and child assemble themselves during interactions and how they allocate and share space and time. Once again, Sander's work helps us comprehend the messiness of development without losing in complexity.

Already in the 1960s, Sander identified the origin of the *assembly grammar* in the first regulation of physiological rhythms, such as sleep-wake cycles. This grammar becomes increasingly characterized by dynamics of behavioral, affective, and cognitive co-regulations throughout development. Louis Sander's visionary contributions paved the way for an entire field of study, suggesting that the seed of our ability to form relationships lies in early interactions between our bodies. More than 40 years later, Ruth Feldman (2007), a prominent researcher in developmental psychobiology, published an article entitled "From biological rhythms to social rhythms: Physiological precursors of mother-infant synchrony."

The system

Sander's contribution was the spark that allowed later scholars—many of whom were his direct or indirect heirs—to describe, in an increasingly detailed and fine way, the different processes of regulation that are at play in parent-child dyads. However, before presenting some of these contributions and their implications for the general goal of this chapter—which is to develop a complexity-inspired perspective on developmental psychobiology—it is necessary to introduce the concept of the *system*.

Von Bertalanffy (1968) defined the system as consisting of components that interact with each other, at different levels, and that are organized into a coherent whole. A living organism is considered a system, particularly an open system, organized into interdependent units that communicate with each other through constant exchanges of matter and information, both within the system itself and outside with the environment, with which it constitutes a larger system. This seamless communication between the living organism and the environment is at the heart of what Sander described; and we can, therefore, consider the parent-child dyad as a system consisting of two sub-systems: the parent and the child. Indeed, it is also a complex system, as endogenous motivations and intentionality can be traced in the system, generally aimed at maintaining and developing the system itself in a coherent way, in space and time. The adaptation of the system is defined by a never-static balance between dynamics that aims at stability and continuity, and dynamics that push toward flexibility and discontinuity. The sense of objectivity that we get when we look at a system—a living organism, a human being, a child, a mother-child dyad—is a sensory illusion linked to the phenotype emerging from the complexity of the interweaving of these messy dynamics.

But what does complexity mean? Why do we say that a system is complex? Of course, it doesn't just mean it's complicated or difficult to explain. Although, certainly, it is difficult to explain a human being. The *complexity theory* (Prigogine & Stengers, 1984) is based on a series of principles of mathematical origin that describe the functioning of complex systems, whether inanimate or life forms. Let's try to give some examples of a complex system to better understand what contributes to its complexity and what are the mechanisms and properties that distinguish it.

First, a dynamic system is capable of self-regulation. This does not mean that the system is closed to external influences; rather, it means exactly the opposite and namely that the system is able to self-regulate the relationship between its components to respond to the stimulations that come from the outside, from the environment. The best example I can give is not related to the parent-child dyad, but it is a particular event—apparently negligible for the themes of this volume— that shocked the ecosystem of an entire continent a few decades ago. Let's move to Australia in the 1930s. Particularly in 1935, when 102 cane toads were introduced to the Far North Queensland region and bred in captivity to solve a problem that Australian rangers had with beetles that were decimating sugarcane plantations. Within two years more than 60,000 specimens were bred in captivity and released. By 1980, these toads would have colonized the entire east coast. While the desired density reduction effect of sugarcane-greedy insects was slow to be observed, soon another and unforeseen problem emerged: these toads were highly poisonous to their predators, and on the Australian continent many potential predators took decades to adapt to their venom, risking extinction. Among them were numerous birds and reptiles. It took many years for predating animals to evolve abilities to resist toad venom or to learn how to hunt them without risking their lives: for example, feeding only on their hind legs, as in the case of the Australian freshwater crocodiles. This ripple effect—largely unpredicted—is known as a *trophic cascade* and it allows one of the principles of dynamical systems to be observed. In fact, the entire ecosystem adapted—self-regulated—in response to the introduction of a new element and did so in unpredictable ways that have implied—more or less quickly—cascade consequences (including mortality) for system components that were already present.

A second distinctive feature of dynamic systems is recursiveness. In other words, when a system does not have to deal with completely new elements from the external environment, its self-regulation processes result in local states of instability, in fluctuations of its parameters within threshold levels that are hardly perceptible from the outside. These fluctuations allow the system to always be in motion, in a dynamic balance that is never definitive. The maintenance of threshold values is governed by distinct parameters designated as degrees of freedom. These parameters ascertain an estimation of system movements preceding the onset of a phase change. The phase transition is externally perceptible, characterized by oscillations surpassing threshold values, leading the system to attain an equilibrium state, distinct from its preceding condition. Naturally, a system in equilibrium tends to avoid persistent phase changes, instead operating within stable threshold levels confined within specific limits. These limits consequently delineate the existence of attractive states—modes of operation to which the system inclines to revert following perturbations inducing parameter oscillations within the threshold levels. Alternatively, a phase transition ensues. Illustrating this concept with an example elucidates its essence. Mood, as commonly understood, tends to be dynamic. Throughout a day, mood fluctuations can be observed in response to various stimuli, such as unexpected encounters, distressing news, accomplishment of goals, work-related or relationship challenges, and so forth. These occurrences typically lead to mood variations that largely fall within the designated threshold values. Conversely, more enduring mood changes may manifest following exposure to acute or repeated traumas capable of inducing oscillations in the parameters of emotional regulation and stress-coping mechanisms beyond the threshold levels stipulated by the degrees of freedom. In such instances, these oscillations can culminate in a phase transition, giving rise to a distinctive mode of emotion regulation that may characterize a stable personality style over an extended period.

A third defining feature of dynamical systems is their inherent unpredictability, contributing to their characterization as non-linear systems. This implies that the trajectories of change, referred to as oscillations beyond the threshold values, are not predetermined but rather influenced by seemingly inconspicuous variables. An illustrative instance of this concept is evident in the subject matter of paintings discussed by Tim Tooney, the trumpeter in the transatlantic Virginian band, within the theatrical monologue Novecento by Baricco (1994, pp. 44–45).

[...] I have always been struck by this matter of paintings. They stay up for years, then nothing happens, but nothing I say, FRAN!, down, they fall. They are there attached to the nail, nobody does anything to them, but at some point, FRAN!, they fall down, like stones. In absolute silence, with everything motionless around, not a fly flying, and they, FRAN! There is no reason. Why at that very moment? We don't know. FRAN! What happens to a nail to make him decide that he can't take it anymore? Does he have a soul, too, poor man? Do you make decisions? He discussed it for a long time with the painting, they were uncertain about what to do, they talked about it every night, for years, then they decided on a date, an hour, a minute, an instant, that, FRAN! Or did they know it from the beginning, the two, it was already all combined, look—I'll give up everything in seven years, for me it's fine, okay then meant for May 13, okay, around six, let's do you minus a painting?, okay, then good night, night. Seven years later, May 13, you are minus a quarter: FRAN! It's something that's better if you don't think about it, otherwise you go crazy. When a painting falls. When you wake up one morning, and you don't love her anymore. When you open the newspaper and read that war has broken out. When you see a train and you think: I have to get out of here. When you look in the mirror and realize that you are old. When, in the middle of the ocean, Novecento looked up from the plate and said to me: [...]

The example concludes here, avoiding spoilers for those who may not have read Baricco's remarkable book. Nevertheless, the central theme is apparent: there is a discernible human inclination to seek meaning by attributing blame to someone or something. Who determined that the painting should fall? In non-linear dynamical systems, the dimension of unpredictability extends to the absence of a culprit, a decision-maker, a puppet master, or a deus ex machina. There exists no superior mechanism or process orchestrating the system—instead, the system, with its intricate complexity, exhibits states of dynamic equilibrium continually shaped by the oscillations or fluctuations of its constituent parts, components, and parameters.

In essence, what we observe in a non-linear dynamic system, exemplified here by a parent-child dyad as the central focus of this text, is the outcome of a continuous negotiation among its components. These components endeavor to sustain coherence within the system, navigating through dynamic disorder. The intricacy of the mother-child dyad as a system stems from the emergence of qualities and properties not directly assignable to either of the two interacting partners. Instead, these characteristics result from the specific and unique way in which each dyad regulates its elective affinities, as elucidated by Sander.

Organization and specificity: The centrality of time

In Freddie Mercury's 1986 song "Time," as he sings, "Time waits for nobody | Time waits for no one | We've got to build this world together | Or we'll have no more future at all," the lyrics underscore a profound truth. It is precisely within this context that we initiate the exploration of how time, especially its rhythmic aspects, plays a crucial role in defining our capacity, as systems, to self-organize and actively contribute to the formation of social relationships and togetherness. At this stage, we should all have a common view of a child within her/his environment: a system comprising various interacting components or parameters, continuously in motion, appearing disordered yet maintaining coherence as development progresses toward increased complexity. Now, the question arises: how do two systems, namely the mother and the child, establish a relationship? In other words, how can we elucidate the integration of these two systems into a singular coherent unit—the dyad—that seemingly operates in a coordinated and integrated manner?

To tackle this inquiry, insights from biology, specifically the concept of organization (Weiss, 1969), can be invoked. Sander contends that organization is "the essential requirement for life to exist." As outlined by Weiss, the organization of living systems pertains to the processes that uphold vital coherence and unity in the organism, guiding its progression toward increased complexity. It represents an emergent and collective property of groups of elements, whose dynamic interrelations interweave them into a higher-order unity. Weiss provides an illustration related to brain functioning. Over a lifetime, the central nervous system must contend with approximately 10^{22} macromolecular constellations, each exhibiting varying degrees of instability and impermanence. While individually, these molecules remain unaware of the vast complexity, collectively, our brain assimilates this information. Consequently, it preserves our sense of unity, individual identity, habits, and memories amidst the continuous and intense modifications within the molecular population.

Furthermore, organization establishes a hierarchical structure, not necessarily in terms of decision-making, but rather in terms of size: smaller components or parameters are encapsulated or integrated into larger or more encompassing ones. As mentioned earlier, in non-linear dynamical systems, there is no central decision-making hierarchy responsible for ensuring unity and cohesion or keeping everything together. So, what provides a unified direction to the movement of a living organism? How do we comprehend the cohesiveness of the organization upon which the possibility of life itself depends? To echo Sander's question: what sustains the dyad's unity while each interactive partner engages in a seamless process aimed at maintaining consistency and ensuring self-regulation?

Weiss incorporated the concept of "specificity" into his theory of the organization of biological systems, denoting the unique encounter between two component systems, each already configured or in configuration with regard to its distinctive properties—referred to by Sander as "prerogatives." Specificity encompasses not only individual behavioral elements, such as acts, gestures, or perceptions, but it also includes the temporal dimension, comprising sequences, timings, rhythms, and co-occurrences of these movements. When a moment of meeting between two systems is characterized by specificity, they resonate with each other, recognizing shared qualities in their respective specificities, such as expressions, rhythms, intensities, behavioral states, speeds, and perceptions. Can we apply this perspective to the parent-child dyad?

This phenomenon has been labeled differently (attunement, synchrony, mirroring) throughout the infant research tradition, each term emphasizing a particular aspect of this dyadic specificity (Provenzi et al., 2018e). In the mother-child dyad, a natural ability to engage in moments of encounter and resonance, akin to those delineated by Weiss, is observable. Consider, for instance, a brief interaction between a parent and their newborn baby just a few hours after birth. As the parent cradles the baby, allowing for eye contact, let's focus on the initial frame: the baby turns their head, looks directly at the parent's face, and in response, the parent widens their eyes and smiles. The baby, reciprocating the smile, briefly closes their eyes and starts reaching the parent's face with their arm. The parent shifts their gaze to the baby's hand and gently grasps one of the tiny fingers, all while maintaining a warm and smiling expression. The baby opens their eyes again and accompanies their smile with a vocalization. This entire sequence, lasting only a few seconds—perhaps three to four seconds—encompasses many of the elements described by Weiss.

First and foremost, we observe the presence of an organization governed by the temporal structure of events. The coordinated sequence of actions and appropriate timing, speed, quality, and intensity of gestures between the two interactive partners not only reveals their individual intentionality but also gives rise to an emergent quality at the dyadic level, which we can refer to as a "relationship." Through repeated exchanges, with variations on the theme, a structure of events is established, representing a shared way of being together and assembling complexity within the non-linear dynamic system comprising the parent and child. This repetition, combined with the potential for variations, provides the child with the opportunity to form expectations, patterns, and predictions about adult behavior and interaction. Simultaneously, the parent becomes increasingly acquainted with their child. This understanding is implicit and dynamic, and Lyons-Ruth (1998) characterized it as a *knowing* rather than a *knowledge*, precisely to underscore the significance of dynamism and the temporal dimension inherent in these forms of learning.

In the context of this temporal structure of events, moments of encounter emerge from the resonance between the distinctive attributes of each partner, encompassing gestures, facial expressions, and intentions. However, it is essential to clarify a significant aspect. When alluding to the intentionality of the child, even in the case of a newborn, it does not imply the same conscious planning associated with adults. For instance, when a child extends their arm toward the parent's face, it is not guided by full conscious planning. Instead, it signifies an implicit intentionality inherently linked to the fundamental nature of being a living organism and consistently experiencing a state of tension or intention within their life context. From this perspective, we can conceptualize the intentionality of the newborn and the preverbal child as a primary activity, generated endogenously by the organism itself and influenced by external stimuli or proprioception. Within this exchange of implicit intentionality, the recognition of mutual specificities can take place. If the caregiver appropriately responds in terms of quality, quantity, and timing to the child's expressions, the child will have the opportunity to recognize their internal states and attribute meaning to them. Initially, this occurs in a biological and implicit manner and later evolves into a psychological and verbally communicable understanding.

Once again, in this brief scenario of a hypothetical interaction, we can observe the pivotal role of time as an element that facilitates the encounter of specificity, ensuring unity and coherence within the dyadic system. Specifically, we underscore the significance of time as rhythmicity. The resonance between the two systems, their mutual recognition, and the sense of togetherness they experience depend on the alignment of their movements in time, synchronized co-occurrence in terms of timing, intensity, and quality of behavioral expressions. Such rhythmicity, particularly bio-rhythmicity, can be considered inherently encoded and embodied in the interaction through the bodies of the two interactive partners (Feldman, 2012). For instance, consider the numerous early parent-child interactions encapsulated and structured around the temporal framework of events, aligned with the child's bio-rhythms. Two notable examples are breastfeeding and the sleep-wake alternation, which inherently follow the natural rhythms of the child's biological processes.

The convergence of bio-rhythmicity in the parent-child system is further supported by what Gergely and Watson (1999) described as a system for detecting contingencies. This system enables the brains of both interactive partners to form implicit modes of learning based on perceiving moments of encounter during their interactions. This mechanism is rooted in a well-known principle in neuroscience, known as Hebb's postulate (1949): when the axon of cell A is close enough to repeatedly or persistently excite cell B, changes occur in both cells, leading to future activations of A being accompanied by concomitant activations of B. Essentially, neurons that are excited together tend to reactivate together in the future. This neural form of associative learning likely underlies the recognition of specificity in terms of quality, timing, and intensity, as described by Weiss. Gergely and Watson expand on this principle by providing a probabilistic description, helping us comprehend how the child's intrinsic motivational system responds to and benefits from the contingent responses of the adult. As previously noted, the child is inherently motivated to process and organize information, seeking regularity, creating expectations, and acting upon them (Beebe & Lachmann, 2003). The system for detecting contingencies would offer the child a probabilistic computation of co-occurrences (temporal, as well as based on intensity and quality) of actions (e.g., baby crying) and outcomes (e.g., consolatory response from the parent). The child's brain would then analyze the probability structure of these contingent correspondences between actions and outcomes—the structure of events. Based on these probabilistic computations, the child's intentional activity can engage in play and experimentation during interactions, with the implicit goal of achieving increasing levels of complexity (development) while maintaining the coherence of the system (self-cohesion).

Inevitable ruptures

Anyone familiar with parenthood or childhood can attest to the understanding that healthy relationships involve periods of togetherness and intermittent separations. These separations allow individuals to tend to their personal needs and navigate the adjustments required to confront unforeseen life events. In this context, Sander's perspectives are illuminating, as he introduces the notion of an "open space." This concept denotes a desirable distance wherein individuals can identify shared similarities and acknowledge differences from others.

This concept was later expanded and investigated by another influential figure in infant research, Ed Tronick. According to Tronick, the adaptation and mutual recognition between the parent (or caregiver) and the child do not follow a linear trajectory. Instead, the primary characteristic of dyadic interaction is the presence of moments of coupling (matching) and decoupling (mismatching). Tronick's concept of matching refers to the mirroring of rhythm, quality, and/or intensity of a behavior or emotional state between the two interactive partners. For instance, a mother smiles at her child who reciprocates with a smile; a father uses a high-pitched voice to accentuate his child's expression of surprise; a child gazes in the direction that the parent is pointing, sharing the attentional focus. These examples illustrate how matching can involve emotional states (such as positive emotionality expressed through smiles), specific qualities of behavioral manifestations (e.g., temporal patterns of surprised facial expressions mirrored in vocal tones), or even socio-cognitive manifestations like paying attention. However, it is crucial to acknowledge that these matching states do not constitute the predominant portion of interaction time. Disconnection states, or mismatches, can be equally prevalent. In fact, Tronick's model of mutual regulation proposes that the primary activity of the dyad entails a continuous negotiation of states of match and mismatch to secure moments of mutual recognition and connection.

To investigate these interactive processes, Tronick developed a specific observational procedure in 1978, known as the still-face procedure (Tronick et al., 1978). In its classic form, this simple observational paradigm lasts for six minutes. During the first two minutes, the mother and baby interact face to face without the use of toys, pacifiers, or other objects—this episode is referred to as Play. In the subsequent two minutes, the mother is instructed to cease all communication with the baby, maintaining a neutral expression while gazing into the baby's eyes without speaking, smiling, touching, or responding to their signals—this episode is termed Still-Face. Finally, the two individuals resume their interaction as they did in the initial episode, and this is referred to as the Reunion episode. This observational paradigm has gained significant traction, with Tronick's original work being cited in over a thousand other publications to date, and numerous studies utilizing the still-face procedure have been conducted in various laboratories involving children with typical development and those at developmental risk (Mesman et al., 2009; Provenzi et al., 2016c). The distinctive feature of this procedure—which renders it an excellent method for assessing dyadic regulatory processes in early childhood (Giusti et al., 2018)—is the opportunity to collect information on the quality of parental behavior during a fundamental interaction (play episode). This episode represents the most common attractor state of the system, yet it also enables observation following a perturbation (reunion episode). During this episode, one can observe how the system restores a previous regulatory state and how this new equilibrium is continually informed and influenced by the perturbation itself.

Furthermore, in the still-face episode, it is possible to observe the regulatory strategies employed by the child to gain the parent's attention, regulate socio-emotional stress, and reestablish a state of balance within their own system. Although the so-called *still-face effect* is often associated with an increase in negative emotionality, avoidance of gaze, and a simultaneous reduction in positive emotionality, attention, and social engagement, this effect should be considered an average outcome across the population. In fact, I am always intrigued by the specific strategies that each child employs during the still-face procedure: some children quickly regain a state of calm through self-regulation strategies; others, while not displaying overt signs of distress, may exhibit slight autonomic indicators of stress, such as hiccups and yawns; some children frequently express smiles and direct their gaze toward the parent, seemingly asking, "Hey, are we still playing?"; and finally, there are children who struggle to regulate their stress response to the point of displaying behavioral and motor dysregulation, characterized by crying, loss of postural control, arching, and persistent difficulty in returning to a state of calm even when the interaction resumes. Even a subtle resistance to resuming interaction in the reunion episode can serve as a noteworthy signal. This effect, known as carry-over, may indicate difficulties in readjustment, but it is normal for the child to exhibit a lingering still effect during the reunion episode for some time. Although the still and carry-over effects are often considered as behavioral outcomes of the child, they are better understood as dyadic adaptations to the disturbance induced by the still-face procedure.

In a 2015 study, Provenzi and colleagues employed a dyadic behavior analysis tool specifically designed to capture the presence of attractor states in the play and reunion episodes within dyads of mothers with 4-month-old babies (Provenzi et al., 2015a). This study demonstrated that the interactive readaptation following the still-face episode involves both interactive partners. During the play episode, the attractive state was characterized by a match between "positive maternal emotionality" and "child involvement," indicating that the dyad was often engaged in activities where the child explored the environment (both physical and relational), and the mother responded with positive affirmations of the child's state. Conversely, in the reunion episode, a dyadic state of mismatch between "maternal positive emotionality" and "child negative emotionality" emerged as the second attractor state. This highlighted how the carry-over effect represents an attempt at spontaneous and unpredictable adaptation of the dyadic system.

Matching, mismatching, and other dance steps

Certainly, matching and mismatching are just two forms that dynamic parent-child interaction can take in its complex and ever-changing nature. Through micro-temporal analysis, researchers have uncovered various forms of dyadic engagement between parent and child. For instance, a review by Leclère and colleagues (2014) explored the concept of synchrony, revealing how scholars interpret this term differently, sometimes using it interchangeably with mutuality, reciprocity, and rhythmicity, and assessing it through various scales. In addition, Stern (1985) delved into the concept of dyadic attunement to emphasize the significance of the structure of mutual events in moment-to-moment interactions at both behavioral and biological levels. Beatrice Beebe and Lachmann (2005) have frequently employed the more descriptive concept of co-regulation, which encompasses processes of self- and hetero-regulation based on the temporal and spatial contingency between the two interacting individuals. Moreover, Fonagy and Target (2006) view the contingent responses of parents to their child's emotional signals as crucial for the acquisition of emotional regulation skills. In addition, marked reflection by the parent, such as exaggerating an expressive dimension like sound element or speed, is considered optimal in promoting the emergence of a distinction between self and other in the child.

To unravel the intricacies of these concepts, often used interchangeably and sometimes with casualness, as well as the diverse dyadic processes that are measured differently across studies, our team conducted a comprehensive review of the literature (Provenzi et al., 2018e). This systematic review brought to light more than 80 articles published in international scientific journals, each defining parent-child interaction according to various constructs: tuning, contingency, coordination, matching, mirroring, mutuality, reciprocity, repair, and synchrony. Among these constructs, mutuality and reciprocity emerged as two overarching and comprehensive meta-theoretical concepts that convey distinct perspectives on interactions: they both imply a dyadic dance in which both partners guide the steps, albeit in different moments and modalities. While they share similarities, slight differences persist between the two: reciprocity places emphasis on the equal relevance, frequency, and intensity of the contributions from the mother and child, while mutuality tends to take a more cautious and conservative approach, recognizing differences in the quantity and quality of the mother and child's involvement.

Moreover, two primary cycles of adjustment can be discerned. In the first cycle, recurrent episodes of contingency—the simultaneous occurrence of simple behaviors—lead to states of coordination that not only require co-occurrence but also demand constancy over time to establish dyadic stability. The ability to sustain coordinated contingency states over time is a prerequisite for the emergence of co-regulation at the level of intentions (tuning and mirroring), which elevates the complexity of the system and facilitates the future occurrence of contingency and coordination states. This process defines a script through which the dyad manages the regulation of synchrony in time and space.

The second cycle aligns with Tronick's model of mutual regulation: alternating states of coupling and disconnection contribute to achieving a level of synchrony, characterized by behavioral and neurobiological co-regulation mechanisms that enhance the likelihood of developing dyadic states of attunement. The repair process, during which parent and child realign in matching states, becomes a pivotal dimension of dyadic regulation. The emergence of decoupling or disconnection provides parent and child access to what Sander termed the "open space"—an interactive realm in which the adjustment parameters of the dyadic system can be renegotiated. The effectiveness of the repair processes—the strategies employed by the parent to help the child regulate states of stress or emotional dysregulation—hinges on the shared understanding and implementation of these strategies between the two interactive partners. Tronick refers to this shared understanding as the "dyadic expansion of consciousness," which represents the primary way human beings learn to regulate their emotions during the early stages of life. It is crucial to highlight that these learnings are predominantly implicit and procedural, taking place at the bodily level and beyond the conscious awareness of the interactive partners. In many ways, within the context of parent-child interaction, learning to regulate emotions resembles learning to ride a bike: it is a process of practice, occasionally experiencing setbacks, but ultimately supporting and encouraging one another. The significance of these negotiations extends beyond emotional regulation and encompasses the creation of meanings about oneself and the world.

Meaning making

As we have observed, when parent and child engage in coordination processes during their interaction, they go beyond merely generating momentary contingencies—they also share affective states and intentions. These mental states reveal implicit and explicit plans in their interactions with the world. In this sense, we can view the parent-child interaction as a space where both mother and child construct meanings about themselves and each other, as well as about their place in the world. This multifaceted process encompasses three key aspects: (1) the development of a structured and stable self in the child; (2) the acquisition of the ability to distinguish between oneself and the other; and (3) the formation and exchange of meanings about oneself and others. These aspects unfold through various dynamic processes and contribute significantly to shaping the parent-child relationship and the child's evolving understanding of their self and the surrounding world. Throughout the first two-three years of life, the child goes through various stages of developing a sense of agency, which involves the perception that their actions have an impact on the physical and relational environment. We can identify at least five levels of increasing complexity in the child's sense of agency.

The first level is the self as a physical agent: the child experiences the world through physical sensations and responses, deriving pleasure or displeasure from the environment. This early self, referred to as the proto-self or core self, is strongly rooted in bodily experiences and proprioception. During the first 3 months, the child seeks perfect contingencies between their movements and environmental responses. However, as interactions progress, they begin to prefer imperfect, but more social contingencies, paving the way for the child's initiative.

Next, the child develops a self as a social agent, based on proto-conversational exchanges even before mastering language skills. These exchanges involve rhythmic, recursive, and predictable interactions with the caregiver, where the child starts to express clear emotional states, moving beyond mere occasional gestures. Around 9 months, the child's sense of agency transforms into a self as a teleological agent. At this stage, they tend to engage in repetitive patterns of action learned through interactions with caregivers. While still lacking a full understanding of mental states in others, they exhibit altruistic behaviors, such as returning objects to adults or completing aborted gestures. As they approach the age of one, the child develops into a mental agent, recognizing themselves in the mirror and building a sense of continuity in time and space. This marks the beginning of an autobiographical self, which will later be expressed through verbal language. Through these various stages of development, the child progressively acquires a deeper and more sophisticated sense of agency, gradually integrating physical, social, and mental aspects of their identity.

Dyadic regulation processes also play a crucial role in facilitating the development of a clear distinction between the self and others. The contingencies detection system, for instance, serves the primary function of promoting the child's self-identification. Actions and perceptions generated by the child result in a probability index close to one, reflecting their perfectly contingent nature (e.g., observing the movement of their own feet). On the other hand, social actions and stimuli from the external world will naturally yield a lower degree of contingency. The contrast in probability structures between endogenous actions and observed/perceived events provides the computational basis for the child to distinguish themselves from the surrounding world.

Studies conducted by Rochat and Striano (2002) using the preferential fixation paradigm have provided substantial evidence for the role of contingencies in promoting the differentiation of the child's self, even in the early months of life. Furthermore, research by Tricia Striano and colleagues (Striano, Henning, & Stahl, 2005) has shown that around 3 months of age, the child shifts from a preference for perfect contingencies (such as exploring their own body) to a preference for high but imperfect contingencies (e.g., imitating the actions of adults). This suggests that the detection of contingencies during parent-child interactions plays a pivotal role in fostering the child's substantial orientation toward the social world right from the earliest stages of life.

Ultimately, interactive exchanges at the behavioral level—encompassing visible manifestations of emotions, cognition, attention, and interaction—serve as the foundation upon which the child, as well as the parent, constructs meanings about the world. These meanings become integrated into the regulatory processes of the system, forming additional parameters and constraints that influence the degrees of freedom within it. In simpler terms, they become life solutions or scripts, habitual strategies of regulation, and complex attractor states that shape subsequent life choices and permeate relationships with others as individuals grow older. For instance, the attachment style can be seen as a specific script that applies to perilous situations: where should I seek a safe base? What can I expect? Do I need to fend for myself? Can I trust someone? Will everything turn out fine? It is fortunate that we are not consistently exposed to dangerous situations, and as we progress through life, there are several motivational systems that guide our present and future behavior.

2 Born to be wired

Sander's research focused on highlighting the distinctive nature of the parent-child relationship during human development. Subsequent studies in infant research have expanded on this concept, emphasizing various dyadic regulatory processes that enable a child's complex development. In addition, the "system" metaphor offers a perspective to view the child within their context, considering multiple levels of regulation. The regulatory processes operate within specific thresholds to maintain consistency over time and space within the system. However, the significance of interactions for humans raises important questions. If individuals are self-organizing systems, what drives them to seek and engage in interactions? What do they aim to achieve through these interactions?

Toward the conclusion of the first chapter, we attempted to address this question by suggesting that interactions provide the child with opportunities for the dyadic expansion of consciousness. In simpler terms, these interactions enable the child to develop and learn how to regulate emotions and construct meaningful relationships with oneself as well as with the physical and social surroundings. This sheds light on the purpose behind our innate desire for interactions with newborns and babies. Moreover, by exploring examples from the field of developmental psychobiology, we can trace the evolution of this attitude. The present chapter presents evidence of our interactive and social nature, illustrating how the interplay between biology and psychology enhances our plastic capacity. This capacity refers to our ability to adapt flexibly to the environment we inhabit. In essence, this underscores the fundamental reasons behind our inherent need to connect with others since birth.

In Taung (South Africa)

In September 2016, I encountered a notable article in *The New Yorker* titled: "Why are babies so dumb, if humans are so smart?" My initial response was one of genuine disappointment. Engaged in the study of early skills in very young children, including preterm infants, I found the title to be wholly inappropriate. Drawing from my understanding of infant research, it's unequivocal that newborns are competent human beings, albeit with limitations attributable to the maturation of their central and peripheral nervous system. Nevertheless, right from the outset, they actively participate in interactions with the human environment surrounding them. Despite

DOI: 10.4324/9781003479314-3

my initial reservations, delving into the article proved to be a remarkably thought-provoking and enlightening experience. Beyond the attention-grabbing title, the underlying question seemed to inquire: "Why can a giraffe, within a few minutes of birth, stand up and take its first steps, and why can a monkey instantly cling to its mother, seeking protection and nourishment, while a human newborn cannot?"

When compared to other mammals, human infants undergo a relatively slower trajectory of brain development. At birth, the brain volume is merely 36% of what is anticipated in adulthood. Over the first few years of postnatal life, the human brain achieves approximately 70% of its adult volume (Knickmeyer et al., 2008). Critical developmental processes, such as axon growth and dendritic arborization, conclude during the initial two years, while synaptogenesis (formation of new synapses) and myelination follow distinct timelines for various brain areas, continuing to mature until adolescence. Furthermore, gliogenesis persists into adulthood (van Dyck and Morrow, 2017). To comprehend the reasons behind this seemingly delayed human brain development, researchers have explored an exceptionally well-preserved infant skull of *Australopithecus africanus*, dating back approximately 2 million years ago, unearthed in Johannesburg, South Africa. Discovered in 1924 by a Northern Lime Company worker in Taung, roughly 400 km west of Johannesburg, this skull's remarkable preservation led to a hypothesis of a brain volume ranging between 380 and 405 cm^3 (Falk and Clarke, 2007). Evidently, traces of cortical convolutions were discernible on the inner walls of the skull. Utilizing modern processing software for skull reconstruction revealed a small triangular shape reminiscent of the frontal suture, commonly known as the anterior fontanelle, present in the human infant's skull (Falk et al., 2012). This opening, the anterior fontanelle, typically closes within the first year of age, between 3 and 9 months after birth. Researchers have postulated that during human evolution, coinciding with the shift from four-legged to two-legged walking, a reconfiguration of the birth canal transpired (Rosenberg & Trevathan, 1995). This reconfiguration resulted from reduced space in the pelvic structure due to the adoption of a bipedal position. Consequently, giving birth to babies with the same brain volume as other mammals became impractical. The emergence of the anterior fontanelle thus signifies an evolutionary adaptation that facilitated the frontal bone plates to slide alongside each other. This, in turn, allowed for continued brain growth post birth, preventing bipedal locomotion from impeding brain development and augmenting neuroplasticity.

The human brain presents distinctive anatomical variations compared to brains of other animal species, emphasizing that intelligence is not solely determined by size; rather, brain architecture plays a pivotal role. For instance, in rodents like the capybara, the cortex weighs about 50 g and contains less than half a billion neurons (Herculano-Houzel et al., 2006). Among primates, the cortex size varies, with macaques having a cortex of approximately 70 g and just under 2 billion neurons, while western gorillas possess a cortex weighing more than 350 g, housing 9 billion neurons (Herculano-Houzel et al., 2011). Elephants, with a cortex weight surpassing 2500 g, harbor 6 billion neurons, with 98% of them located in the cerebellum (Herculano-Houzel et al., 2014). In contrast, humans exhibit a cerebral cortex weighing around 1200 g populated by approximately 16 billion neurons. This accentuates the privileged role of the cortex, which has undergone the most significant increase in volume throughout evolution. The specificity and significance of neuroplasticity for the human brain are further emphasized by a 2015 study revealing that human brains are less genetically heritable and more influenced by environmental experiences, in comparison to other primates like chimpanzees. Gomez-Robles and colleagues (2016) demonstrated that while brain size may be highly heritable in both species, the organization of the cerebral cortex, especially in areas linked to cognitive function, is much less constrained by genetic control in humans.

Exposure to parental care during the period of ongoing brain maturation provides the human brain with heightened flexibility and plasticity. This adaptability enables the brain to continue growing postnatally, utilizing experiences from the external environment beyond the physiological confines of the uterus (Trevathan, 2015). Piantadosi and Kidd (2016) discovered a substantial positive association between the amount of time spent in care and measures of offspring intelligence in primates. Notably, primates like pongos and gorillas, engaging in more caring activities with adult conspecifics, exhibit higher intelligence indices compared to other primates like lemurs or talapoins, where care time is diminished. The "Machiavellian intelligence" hypothesis (Byrne & Whiten, 1988) further underscores the connection between competitive social interactions and brain development in hominids. Overcoming challenges in a resource-constrained social context has played a role in the evolutionary success of the human brain, contributing to its specific specialization and growth of the cerebral cortex. In essence, the unique architecture of the human brain, coupled with its adaptability to external experiences and the intricacies of social interactions, significantly contributes to its remarkable cognitive abilities.

In utero

Human brain maturation extends well beyond the conclusion of gestation, and there is growing evidence suggesting the occurrence of interactive exchanges and biological communications even before birth. Over the past four decades, interest in the fetus within sociocultural contexts has experienced a substantial surge, resulting in notable shifts in societal perceptions of the fetus. Since the 1980s, the presence of pregnant women and fetuses in various media outlets has increased, fostering curiosity about prenatal life. Concurrently, scientific curiosity in fetal skills, especially early sensitivity to environmental stimuli, has steadily expanded since the 1990s.

Research indicates that even at 36 weeks of gestational age, fetuses can discriminate between maternal voices recorded by devices and live voices produced by mothers, as evidenced by the number of fetal movements (Hepper et al., 1996). The fetus also responds to tactile stimulations on the abdomen from both the mother and father, exhibiting distinct movement patterns based on the source of touch (Marx & Nagy, 2015).

Moreover, variations in physiological regulation, such as heart rhythm, may reflect the fetus's response to external stimulation. Fetal heart rate patterns in the third trimester exhibit circadian-based periodicity, with correlations between fetal and maternal heart rhythms (Stark et al., 1999). The maternal voice influences fetal heart rate, with an increase observed when exposed to the mother's voice and a deceleration when exposed to less familiar voices (Kisilevsky et al., 2003, 2009). In addition, maternal stress states can affect fetal heart rate and movement (Di Pietro et al., 2003). The mutual perception of physiological and neurobehavioral rhythms between mother and fetus is deemed crucial for the emergence of proto-conversational and intersubjective rhythms, even before birth, fostered by the intrauterine maternal-fetal connection. Neuroimaging tools have unveiled cerebral responses in the fetus to various stimulations, including distinct activations in response to the maternal voice, suggesting early involvement in sound processing and emotional stimuli (Jardri et al., 2012; Abrams et al., 2016). Furthermore, fetal exposure to specific melodies during pregnancy may leave neural memory traces, leading to enhanced responses to the same sound stimulations after birth (Partanen et al., 2013).

While these studies offer intriguing insights into early fetal development and the potential impact of prenatal experiences on socio-emotional and socio-cognitive trajectories, they come with limitations. Small sample sizes and partially documented methods complicate comparisons, and the presence of confounding variables in pregnancy further challenges direct access to fetal responses, whether neurophysiological or behavioral. Nevertheless, advances in scientific methods and technologies are facilitating more reliable and integrated studies, enabling a better understanding of how early life experiences, even before birth, shape the developmental trajectories of human beings in the social and emotional realms.

Just like me

After birth, our inherent inclination for interaction becomes increasingly apparent. Neonatal imitation emerges as a potent learning mechanism for newborns, primarily owing to two factors. First, it is facilitated by motor repertoires inherently present from birth, such as tongue protrusion. Second, adults frequently engage in imitative interactions with newborns during the early hours and days of postnatal life, popularizing this mirroring game. Neonatal imitation stands out as one of the most well-acknowledged manifestations of the human predisposition to connect. Intriguingly, it represents an early form of social interaction that surfaces before other manifestations of socio-emotional or socio-cognitive learning. When an adult captures the attention of a newborn positioned in front of them and consistently performs specific facial actions, vocalizations, and movements, the newborn endeavors to imitate these actions in immediate and progressively refined approximations. Andrew Meltzoff, a distinguished scholar of neonatal imitation, has amassed abundant evidence demonstrating that within a few days and hours after birth, newborns exhibit imitative actions in response to adults (Meltzoff & Moore, 1977). This evidence has been extensively corroborated and replicated in scientific literature (Simpson et al., 2014), indicating that the capacity for imitation is inherently intermodal or supra-modal, independent of specific sensory channels (Ferrari et al., 2003; Simpson et al., 2014). Aligned with the immediate nature of neonatal imitation, Ferrari suggests that sensory information, primarily visual in the case of imitation, can directly reach motor regions, activating gestural coordination corresponding to visual mapping.

It is well-established that in children, imitative abilities and a preference for faces, particularly the eyes, are correlated with subsequent socio-emotional and socio-cognitive development (Brenna et al., 2013; Turati et al., 2002). Employing eye-tracking technology, Ferrari and colleagues (Paukner et al., 2014) conducted a study on macaques, just a few days old, exposing them to an animated avatar displaying a motionless face followed by affiliative behaviors (lip-smacking or tongue protrusion). The frequency of spontaneous imitations in the macaques predicted their subsequent social attention to the avatar's eyes. This suggests that early engagement in imitative interactions may foster additional socio-cognitive abilities, such as social attention, during early childhood.

Imitation is commonly regarded as one of the precursors of secondary intersubjectivity, encompassing socio-cognitive skills that emerge around the 9th month of life, including declarative pointing and attention triangulation. These skills signal the development of a rudimentary theory of mind. Meltzoff proposes that imitation leads to a progressive transition from an initial equivalence between self and others to the child decoding and memorizing the relational repetitions between their own behavior and relative mental states. Consequently, the child recognizes that other social subjects act similarly, possess similar intentional mental states, and have a mind. This recognition, termed as "like-me-ness" by Meltzoff and Moore (1977), prompts the question of how it unfolds.

One widely held possibility is embodied simulation (Gallese, 2007), which is based on the discovery of mirror neurons in the ventral premotor cortex of monkeys. These mirror neurons have special visuomotor properties, activating when the animal performs an action or observes a similar action being performed by another. In humans, a mirror neuron system has been documented, responsive to the observation of targeted actions performed with the hands, face, mouth, and even sounds associated with motor representation of an action. In children, electroencephalography (EEG) methods, particularly the mu rhythm, have been used to study the response of the mirror neuron system (Filippi et al., 2016). The mu rhythm, a type of sensorimotor rhythmicity, has been identified as a potential reliable neurophysiological marker of mirror system activity. Studies on monkeys have shown a significant correlation between mirror neuron activity and the suppression of the mu rhythm in the EEG. In children, desynchronization of the mu rhythm has been observed in response to the observation of finalized movements of the hands at the age of 7 months.

Investigating the mirror system activity in preterm infants can offer interesting insights. Preterm birth poses early risks for child development due to factors like separation from parents, incubator treatment, and exposure to a stressful environment. This could impact the typical development of the mirror system, which begins during fetal age and involves visual explorations of one's body in motion. A recent Italian study (Montirosso et al., 2019) compared full-term and preterm babies at 14 months of age during imitation tasks. Both groups showed suppression of the mu rhythm during the execution of the action (imitation). However, different patterns emerged during the observation of the action, suggesting atypical processing of observed targeted actions in preterm infants. Early parent-child physical contact, facilitated by interventions like skin-to-skin contact, is crucial for the well-being of preterm infants. In healthy children born at term, desynchronization of the mu rhythm is observed in response to received or observed caresses (Addabbo et al., 2020). Future research will investigate whether facilitating early skin-to-skin contact in preterm infants can protect and foster the development of the mirror system in conditions of evolutionary risk.

Giant cells

Mirror neurons are not the sole population of nerve cells in our brain associated with our inherent inclination for social connections. Another type of nerve cells, known as von Economo neurons (VEN), are present in various species, including humans, chimpanzees, bonobos, orangutans, macaques, whales, cetaceans, and elephants, all of which share a common characteristic of possessing a high brain volume (Allman et al., 2005). These neurons have been historically characterized as large spindle cells, with the first descriptions dating back to the late nineteenth century (see for example, Betz (1881)). The most detailed description of the morphology and distribution of these spindle cells in the cerebral cortex was provided by Constantin von Bursar in 1926 (Seeley et al., 2012): cells that showed a typical spindle shape and an unusual length, oriented perpendicular to the pial surface, which possessed apical and basal dendrites as wide as the cell body. Already von Economo also anticipated the peculiar distribution of this type of cells in the fifth layer of two specific brain regions: the anterior cingulate cortex and the anterior insula. Only recently, Allman (Allman et al., 2005) renamed these neurons as VEN to prevent them from being confused with other fusiform nerve cells present in other layers of the cortex. In fact, unlike pyramidal nerve cells, these neurons have a single dendrite.

The total number of these cells is vastly greater in humans than in other primates: it is estimated that about 200,000 VENs are present in adult humans, while there seem to be less than 10,000 present in non-human primates (Allman et al., 2005). Therefore, there seems to be a phylogenetic specialization that dates back to about 13 or 15 million years ago in hominids and that has proliferated widely in the evolutionary line of man. Ontogenetically, VENs have been observed at 35 weeks of gestational age in higher densities in the right anterior insula, but not before (Allman et al., 2010). The postnatal increase in the number of VEN peaks at 8 months of age, with an increase of about 800% in the right anterior insula. The number of these neurons stabilizes, reaching a population density typical of adulthood around four years of age. At this age, VENs seem to be more in number—about 30%—in the right hemisphere, which has led to the hypothesis that they are particularly involved in the processing of social emotions, especially since their migration or specialization in the related brain areas occurs at an early stage of development where the quality of parental care and parent-child interaction plays a crucial role. It should also be noted that both the anterior cingulate cortex and the anterior insula are activated in response to images of a loved one, which makes these regions particularly involved in the formation and maintenance of affective bonding (Bartels and Zeki, 2004). In addition, the same regions are activated in response to situations characterized by errors or social ruptures. Examples can be social contexts characterized by the violation of expectations such as resentment (Sanfey et al., 2003), deception (Spence et al., 2001), embarrassment (Berthoz et al., 2002), and guilt (Shin et al., 2000). Most commonly, both of these regions are associated with empathic responses to the suffering of others (Walter, 2012); in mothers, the anterior insula—particularly in the right hemisphere—seems to be specifically responsive to the baby's crying (Riem et al., 2011).

Anatomical studies that have documented the presence of VENs in humans, large primates, cetaceans, and elephants seem to suggest a comparable cortical distribution. However, the ratio of VEN/pyramidal neurons is much higher in humans than in some species of elephants in the anterior insula (Hakeem et al., 2009) and cetaceans in the anterior cingulate cortex (Butti et al., 2009). The presence of these nerve cells in similar cortical areas suggests that phylogenetically the VENs specialized independently under the pressure of comparable evolutionary pressures in animals with very different evolutionary histories. Therefore, although they were initially discussed as "the neurons that make us human" (Butti et al., 2011), one could instead think of VENs as neurons that support and predispose different animals to specific behaviors related to survival in extended social contexts. In particular, Allman and colleagues (2010) hypothesized that the observed peak around 8 months and the stabilization observed around four years in the proliferation of VEN cells signal their involvement in the early emergence of socio-cognitive skills that are precursors of the theory of mind (Craig, 2009).

Given their potential involvement in behavior and social cognition, some studies have focused on VENs in the brains of subjects with different neuropsychiatric disorders. A reduced number of VENs was found in subjects suffering from dementia and in whom the capacity for self-control and the inhibition of inadequate social behaviors was deficient (Seeley et al., 2007). Patients with agenesis of the corpus callosum—who show difficulties in processing social stimuli (Paul et al., 2004)—present with a reduced number of VENs in the anterior cingulate cortex and insula (Kaufman et al., 2008). Both the anterior cingulate cortex and the insula participate in the neural processes underlying the integration of emotional stimuli and bodily proprioception (Craig, 2009), processes that seem to be compromised in subjects with autism spectrum syndrome (Ben Shalom et al., 2006). Santos and colleagues (2011) analyzed the postmortem brain tissues of four patients diagnosed with autism and three matched controls for sex and age and quantified the number of VENs and pyramidal neurons in the fifth layer of the insula. In the brains of patients with autism, the ratio of VEN to pyramidal neurons was significantly higher than in the control brains. These results would apparently suggest that in subjects with autism a hyper-production of VEN neurons may occur, possibly as a result of altered migration or apoptosis processes. Their functional correlate is difficult to speculate, although the authors hypothesized that they could underlie increased enteroception of patients with autism. In general, however, it remains complex to study VEN in depth and their functional implications in humans, given the impossibility of conducting invasive studies. The studies conducted so far have been conducted on postmortem brain tissue samples and therefore are purely histological and anatomical, not functional. These limitations make it very difficult to obtain direct evidence of the functional role of these neurons in humans.

The skin as a place to meet

If we examine development from the perspective of a non-linear dynamic system, it becomes incongruent to propose the existence of a homunculus or hierarchical processes responsible for top-down control of our organism, as a dynamic system does not allow for such hierarchies. Despite the unique characteristics of the human brain, which has evolved into the most powerful tool, peripheral processes can also play a fundamental role in regulating the parameters of a living system. Specifically, certain sensory fibers in the periphery of our body seem to have a special role in perceiving and regulating social and affiliative exchanges.

Physical and tactile exchanges between parent and child are widespread and play a crucial role in the social, emotional, and physical development of the child. Maternal touch during the neonatal period, for example, is associated with a greater ability of the child to exhibit appropriate stress-regulating strategies in children of depressed mothers (Sharp et al., 2015). Early physical contact between mother and baby after birth reduces the baby's crying and stress manifestations, supporting the newborn's adaptation to the post-natal living environment (Winberg, 2005). The patterns of tactile behaviors exhibited by the mother shortly after birth tend to remain relatively stable in the following months, indicating identifiable styles and recursive modalities of physical contact that contribute to co-regulation within a given dyad (Mercuri et al., 2019).

Studies conducted by Dale Stack of Concordia University in Montreal, Canada, using a modified version of the still-face procedure, demonstrate that the presence of physical contact between the mother and baby significantly reduces the child's stress response (Jean et al., 2014). Furthermore, touch facilitates the perception and recognition of one's own body, contributing to multisensory integration in space and time (Filippetti et al., 2015, 2013). Infants as young as 4 months show high sensitivity to spatio-temporal correspondences between tactile and auditory stimulations (Thomas et al., 2018). This suggests that somatic experience in the parent-child relationship is essential for effective multisensory integration.

During early development, the child's bodily self is a multisensory entity in search of coherence, regulation, and integration between various sensory stimuli. Through self-exploration and perception of sensorimotor contingencies, the child begins to distinguish perceptions that concern themselves (perfect contingencies) from those related to the social-interactive world (imperfect contingencies)

(Gergely & Watson, 1999). The insula, a cortical region that develops early and shows mature characteristics as early as 27 weeks of gestational age, plays a major role in integrating external sensory signals and signals from one's own body into a cohesive scheme.

Touch, especially affective touch conveyed by C-tactile fibers (CT fibers), plays a significant role in this integration process. CT fibers are non-myelinated and mechano-sensitive fibers that respond to gentle touch with a speed of 1–10 cm/s, similar to the speed of a caress. They reach the posterior dorsal nucleus of the insula and brain circuits involved in perception and social processing. Maternal social touch, which activates CT fibers, seems to be specifically responsive to this type of slow, affectionate touch (Jönsson et al., 2018; Tuulari et al., 2019). The response of CT fibers to affective touch is already present in preterm infants and may modulate the neurophysiological response of premature infants to stress (Tuulari et al., 2017; Manzotti et al., 2019). Maternal gentle touch can also support social attentional processes in children with disabilities and psychomotor impairment (Provenzi et al., 2020e). Thus, physical proximity and affective touch between parent and child contribute to an integrated strategy of processing sensory data from both internal and external sources, facilitating dyadic co-regulation. This embodied process plays a fundamental role in the child's development and the establishment of an integrated sense of their bodily self (Montirosso & McGlone, 2020).

Conclusions

In this chapter, we have delved into the evidence of psychobiological predispositions that enable us to connect with others from the earliest stages of life, perhaps even during fetal development. These lines of investigation vary in their level of advancement. While VENs remain relatively unexplored, research on the mirror system in humans has significantly expanded and it markedly contributes to our understanding of early socio-emotional and socio-cognitive functioning in both typical and at-risk development. Our body is not merely a vessel for the mind to connect with others; instead, it actively participates in socio-relational interactions. The ability to draw from our bodily experiences during interactive exchanges allows us to develop a coherent, stable, and continuous sense of self in both time and space.

3 The assembly grammar

In 1998, Lear asserted that "Freud is dead. He died in 1939 [...] It is important not to fixate on him, like certain tenacious symptoms: it makes no sense either to idolize him or to denigrate him." However, I find some passages in Sigmund Freud's writings still astonishing due to the clarity and ambition of the Viennese doctor—not only the father of psychoanalysis but also a skilled neurologist—in attempting to integrate the limited biological knowledge of his time into a comprehensive model of the human psyche that explained its development, symptoms, and psychological suffering. In his work "Beyond the Pleasure Principle" (Freud, 1920), he states, "[...] We must expect [from physiology and chemistry] the most surprising explanations; We cannot therefore guess what answers they will give, in a few decades' time, to the problems we have posed. Perhaps these answers will be such as to bring down the whole artificial edifice of our hypotheses."

Certainly, Freud exhibited a prominent element of reductionism or energy empiricism—a characteristic reflective of the scientific culture of his era (Assoun, 1988). This inclination toward reductionism led to a drift in his thinking, as evident from his statement: "We must remember that all the psychological notions that we are gradually formulating must one day be based on an organic substrate" (Freud, 1914). Today, the epistemological framework of non-linear dynamical systems provides us with a suitable context for exploring the assumptions and neurobiological contributions to the development of the self. It allows us to maintain a heightened awareness of the interplay between different parameters and processes involved, while avoiding the temptation to simplify the complexity of our self to mere molecular components.

In this chapter, I aim to explore the fundamental processes that connect the functioning of our central nervous system to two essential aspects of our daily lives: learning and remembering. These abilities are inherently woven into the fabric of our brains and, more broadly, our bodies. As we discussed in Chapter 2, not only are we born to be connected, but we are also innately predisposed to learn. Hebb's axiom, which outlines a basic form of associative learning based on detecting contingencies, underscores the inevitability of gaining experiences. On the one hand, we possess a complex biological system that is inherently geared toward facilitating learning. On the other hand, this very system is receptive to external influences and experiences, allowing them to shape its development. Through processes of

DOI: 10.4324/9781003479314-4

neurobehavioral plasticity, experiences play a crucial role in fostering the most optimal adaptation and striking a balance between the system's constraints and degrees of freedom, complexity, and cohesion. This leads us to a fundamental question: to what extent is our capacity for learning rooted in innate assumptions, and how much of it depends on our interactions with the world, making experiences an integral part of ourselves?

To select and to instruct

The inquiry into innatism is a longstanding subject in epistemology and the scientific disciplines dedicated to examining and comprehending human nature, especially its methods of learning. In ethology, there has been a longstanding reliance on innatist concepts to elucidate the origin and significance of certain behaviors observed in all members of a given species, irrespective of their shared developmental context. Ethologists have categorized behaviors like parental care and responses to predator-related warning signals as innate. Essentially, these behaviors were thought to manifest independently of an individual's learning and experiences, ingrained in the inherent heritage of every member.

In the realm of psychology, particularly mid-twentieth-century North American psychology, behaviorism heavily influenced developmental theory. According to behaviorism, experience and learning played a pivotal role in shaping an individual's behavior. The black box theory posited that, apart from central processing, individuals could be comprehended based on learning derived from experiences, especially those associated with environmental conditioning. The behaviorists asserted that, regardless of the species, learning was the sole mechanism at play. Consequently, the extrapolation of findings from animal models, exemplified by the well-known superstitious pigeon experiment, to human functioning was considered a valid scientific method. This perspective gained popularity and endured for years until Noam Chomsky, in the 1950s, challenged the assumption of a universal learning capacity. Chomsky's theory of a universal grammar, inherently encoded in the human genome, disrupted the meta-theoretical framework of behaviorism. While giving rise to a new psychological paradigm, cognitivism, Chomsky introduced a novel concept of innatism associated with advanced mental functions, particularly language.

The notion of universal learning rules, presumed to be applicable across various animal species, faced a new challenge with the identification of biological constraints—system limitations encoded at the genetic level—that profoundly influence learning processes. In the 1980s, psychologist Steven Pinker proposed a theory that sought to integrate Chomsky's innatism with evolutionary concepts. Pinker examined pidgin, a form of verbal communication adopted by individuals speaking different languages but needing to establish a common language for reasons related to coexistence. This phenomenon is typically observed in ethnic minorities experiencing migration, colonization, or labor relations, often spanning a single generation or a local group. Examples of pidgin include chinglish, a linguistic blend of English and Chinese in Southeast Asia, and Fanakalo, rooted in Zulu and serving as a lingua franca in South African mines and other African regions. Pinker observed that the

creolization of pidgin over generations imparts greater grammatical structure, leading him to posit that this process hinges on an innate nature underlying the assemb-
lative grammar of language. Similar innatist perspectives have been extended to the
development of various human skills, encompassing not only cognitive abilities but
also universal aspects of emotional expressiveness. However, while emotions may
exhibit largely universal and cross-cultural forms of expression, their significance
and interpretation emerge within interactive social contexts.

The interest in studying the innate character of human cognitive skills has given
rise to the so-called evolutionary psychology, which aims to understand how certain
skills that appear innate have been selected and sculpted by processes of natural
selection. As LeDoux (2000) reports, the theory of evolutionary psychology offers
intriguing explanations that are difficult to falsify. However, Gould himself defined
the explanations of evolutionary psychologists as "just-so stories," meaning partially unverifiable narrative explanations. Gould also coined the term "spandrel"—
literally, a plume—to label traits that may have been selected in an ancillary way
through the selection of other adaptive traits, rather than conferring an adaptive
advantage on a species. In contrast, traits that are not directly selected but that end
up conferring a new evolutionary advantage are defined by Gould as *exaptations*
(Gould & Vrba, 1982)—literally, we could translate it as exact. This is a way in
which Gould intends to avoid confusing the adaptive and utilitarian character of a
trait at a given moment in phylogenetic history with its evolutionary origins. Gould
directly quotes Darwin (1859) to support this point: "It is interesting to contemplate a lush plain, lined with many plants of various kinds, with birds singing in the
bushes, with various insects buzzing around, and with worms crawling in the moist
soil, and to think that all these forms, so elaborately constructed, so different from
each other, and dependent on each other in such a complex way, were produced by
laws that act around us." It is ironic that Darwin himself—the banner of the most
extreme current of the supporters of evolution and natural selection—was, in fact,
perhaps the first of the so-called pluralists, that is, of that current that considers
natural selection a fundamental principle acting in a context of unpredictable contingencies. In essence, exaptations are characteristics that have enhanced our adaptive abilities at some point in evolutionary history but were not originally intended
to fulfill that exact purpose. Examples of exaptations include the feathers of birds,
which initially evolved to favor thermal regulation and only later came to support
the ability to fly. Another example is the dog's gesture of licking. While the ancestral purpose of licking in canines might have been primarily related to grooming,
cleaning, and showing submission, it has evolved to serve additional functions.
In modern domesticated dogs, licking can have various meanings and functions.
Dogs often lick their owners or other dogs as a form of social bonding, communication, or even to seek attention and affection. The concept of exaptation also
suggests that once a complex system is assembled, emergent functions—therefore,
unexpected ones—can be developed, forged, selected, strengthened, or lost.

Today, the debate on innatism is no longer so much a question of the preex-
istence of skills and knowledge, but rather the possession of meta-competences
that configure a predisposition to specific forms of learning. A widely accepted

idea is that there are genetic programs that set up the system, helping to make all members of a species highly similar or probabilistically similar, while also offering each member the possibility to interact with the genetic program based on their own experiences, leading to the development of unique phenotypes with different degrees of probability. In fact, genetics offers living systems a common biological substrate for a series of associative or conditioned learnings, such as habituation and sensitization, which will be discussed later in the chapter. These molecular mechanisms enable all the structures of the central nervous system to learn and modify their functioning, including their conformation and connections, based on experiences. From this point of view, it is possible to hypothesize that learning is a characteristic inherent in the functioning of our nerve cells—as a whole—and does not represent a specific function selected by evolution. Learning becomes a fundamental condition of life for our system as it is configured.

If this proposition holds true, the inquiry transitions to the neuroscientific realm, focusing on the configurations of specific brain circuits, not so much within the scope of phylogeny but rather during ontogenesis. In this context, a dichotomy emerges between proponents of "selection" and "education." Advocates of neo-Darwinism, exemplified by figures like Changeux, Mehler, and Palmarini, posit that learning doesn't simply occur through the assimilation of environmental information by the body, but rather via the internal selection of brain connections. Knowledge, in this view, results from a process of reduction, aligning with the renowned design principle of Dieter Rams: "less is more." Neural Darwinism, a product of this perspective, has, through Edelman's contributions, offered profound insights into the role of experience in governing the synaptic connections within our brain, adhering to the "use-it-or-lose-it" principle. Edelman contends that environmental influences play a pivotal role by contributing to the selection of synapses that activate most frequently and significantly, thus promoting the consolidation of specific patterns of neural activity while disregarding others. On the opposing side of this debate, constructivists or instructionists assert that brain development is guided by instructions derived from experience, emphasizing the comprehensive role of brain plasticity rather than the supremacy of the cortex. Examples presented by these contrasting viewpoints often pertain to diverse learning contexts: the focus on high cognitive functions within the neo-Darwinian innatist perspective and the consideration of perceptual and sensory learning within the constructivist framework. According to the latter, genes establish a fundamental structure, yet it is the interaction with life events and immersion in experiences lacking continuity that shapes epigenetic processes of synaptic instruction and selection. The exploration of brain plasticity simultaneously constitutes a fundamental aspect of the constructivist—and subsequently neuro-constructivist—approach and a field of inquiry that has significantly advanced owing to these theoretical frameworks.

Not in a vacuum

A prominent place in this debate—at least from the point of view of developmental psychology—deserves the approach of neo-constructivism and the role of

Annette Karmiloff-Smith within the epistemology of development in psychology. Karmiloff-Smith moves from a critique of the constructionist models of the adult, suggesting that they are inadequate to understand development: "[...] Although I believe that Piaget was wrong to theorize the stages (Karmiloff-Smith, 1992), his was undoubtedly a halfway epistemological position, which emphasized that innatism was a theoretical excuse [...] and how the search for an initial stage was a static challenge that could stimulate an endless regression since, in the deepest sense, there is no beginning [...]."

Although the notion that the brain comprises relatively independent functioning modules may be applicable to the central nervous system of the adult individual, which is capable of new domain-specific acquisitions even in the presence of focal damage, this conception cannot be extended to the child and development. In other words, the hypothesis that the modular nature of mind cannot be assumed (Fodor, 1983) is of an innate character or innately specified and determined. This assumption would ignore—according to Karmiloff-Smith—the real processes through which the human brain develops at the ontogenetic level (Karmiloff-Smith, 2009). His position resonates with Spencer's regarding intelligence as a process—rather than a static competence—that comes to be configured through the reciprocal influence in time and space of factors such as genetics, neurology, cognition, behavior, environment (Karmiloff-Smith, 1998). As already mentioned in Chapter 1, time plays a fundamental role in this complex plot.

Karmiloff-Smith (Elman et al., 1996), on the other hand, proposes a position that represents a path of integration: it does not argue that the child's brain is a universal learning device, nor that there are preordained and domain-specific modules. Rather, it proposes that subtle regional differences in the type, density, and orientation of neurons, neurotransmitters, discharge thresholds, myelination rate, gray to white matter ratio, and so on, determine that certain brain circuits are somehow more relevant in processing specific information than others (Karmiloff-Smith, 1998). From an evolutionary standpoint, numerous brain regions might possess the capacity to analyze various forms of information. However, with recurring exposure to the environment and accumulated experience, certain circuits tend to develop a heightened capability to finely and specifically analyze particular types of information, excluding others. This experientially driven process is notably observable in brain areas responsive to human faces, demonstrating specialization during childhood. Another illustrative example, as suggested by Karmiloff-Smith, involves the primary visual cortex, which exhibits responsiveness and processing capability for tactile inputs (as observed in Braille) or auditory inputs (Sadato et al., 1998). Hence, it is the alterations that transpire over time due to life experiences that serve as guiding forces in development.

Karmiloff-Smith's neuro-constructivist approach repositions the role of experience in human development, particularly during early stages, as a central factor. The examination of face processing in individuals with Williams syndrome serves as a prototypical and frequently cited case by Karmiloff-Smith to reinforce her stance on the interplay of genes and environment in early development. While some authors have proposed the existence of an intact face processing module in individuals with Williams syndrome (Bellugi et al., 1994), Karmiloff-Smith underscores that the processing pathways employed by these individuals to achieve similar facial processing performance to those with typical development are distinct. Behavioral and neurophysiological evidence indicates that individuals with typical development employ configural strategies in face processing, whereas those with Williams syndrome use approaches linked to individual features or holistic elaborations independent of facial architecture (Grice et al., 2003). Rather than displaying an intact specialization for faces and a specific difficulty in spatial processing, the brains of individuals with Williams syndrome fail to modularize face processing during development.

Language development in Williams syndrome provides insights into the impact of experience on the unfolding cognitive phenotype of the child. Infants with Williams syndrome exhibit specific limitations in hand movement, babbling, and verbal flow segmentation, relying more on perceptual stimuli than linguistic labels for identifying new objects (Masataka, 2001; Nazzi et al., 2003, 2005). Early categorization abilities appear partially compromised, and certain socio-cognitive skills, including pointing and following the interlocutor's gaze, exhibit distinctive delays. These specific alterations, extending beyond the realm of language, contribute to cascading effects that can result in pronounced linguistic and cognitive deficits. However, the genetic underpinnings of language, long upheld—especially following studies on the KE family, where only members with language disorders shared a FOXP2 gene mutation—are now under scrutiny and challenge. Contrary to earlier suggestions by Pinker (1994) and Wexler (1996), the genetic origin appears less direct, and notably, the association with the FOXP2 gene mutation remains unconfirmed in subsequent studies.

More broadly, neuro-constructivism posits that a straightforward one-to-one mapping between genes and cognitive developmental outcomes is implausible. Genes are likely to contribute to system constraints at a more general level, such as defining specific time windows, neuronal migration, size, and other characteristics of nerve cells. While different configurations of these parameters may render a circuit more or less relevant to a particular domain, it is the trajectory of development—shaped by the subject's experiences in space and time, along with the iterative processing of these experiences—that decisively contributes to the emergence of specific and modularized domains observable in the adult model. It is challenging not to concur with Karmiloff-Smith (2006) assertion that "if one thinks from a truly evolutionary theoretical framework, it becomes immediate to consider how even small asynchronies or early deviations in development can exert enormous and cascading effects on the phenotype." Consequently, it is no longer tenable to conceptualize child development as a genetically predetermined process, following maturation stages and independent of experience and environmental influences. The child—indeed, every organism—is not a system that operates in isolation.

Frog thighs and hearts

The stone age of our conceptions of the brain ends at the beginning of the twentieth century with the works of Santiago Ramón y Cajal, the theory of the neuron, and the birth of what we now call neuroscience. In fact, the idea that the cell was the primary unit of functioning in eukaryotic organisms had already been advanced—in the plant world—in 1837 by Schleiden, a German botanist. It was Schwann in 1838 who extended this hypothesis to animals. The presence of "filaments" accompanying brain cells was an element that, for many years, divided scholars and anatomists about the applicability of the cellular theory to the brain. Lattice and cellular theory resisted as valid options until the publication of Ramón y Cajal's work.

The narrative is well-documented: utilizing brain staining techniques developed by Camillo Golgi—a proponent of the reticular theory—which facilitated improved visualization of nerve fibers and the microscopic structure of the brain, Ramón y Cajal, in 1891, successfully identified distinct cells surrounded by membranes. He postulated that these cells communicated with each other only at specific points of contact. Interestingly, despite Golgi remaining a staunch advocate of the reticular theory, both researchers were jointly awarded the Nobel Prize in 1906 for the discovery of the neuron. It was Wilhelm Waldeyer, a follower of Ramón y Cajal and a staunch supporter of the cellular theory of the brain, who coined the term "neurons" to describe these cells in 1891. In truth, as early as 1883, another Viennese scholar, Sigmund Freud, had almost prefigured the times. In his studies on the nervous system of fish and lobsters, Freud suggested that brain cells were not part of an indistinct network but were instead separated from each other. This concept resurfaced in Freud's 1895 work, "Project for a Psychology," an unfinished manuscript in which he, among other things, offered an ante-litteram description of a somatosensory resonance system reminiscent of the mirror neurons described a century later by Rizzolatti and colleagues. In the Freud Project, he also introduced the concept of "contact barriers" to describe the way in which nerve cells communicate with each other; in 1897, Sherrington would term them "synapses."

Ramón y Cajal is commemorated as the progenitor of neuron studies, notably for his subsequent investigations wherein he deduced dynamic properties of nerve cells from static brain images, elucidating the principle of dynamic polarization. This principle posits that electrical signals traverse the neuron from the receiving pole through the dendrites, cell body, and axon to the subsequent synaptic termination. A consequential outcome of this research was the ability to explore brain activity within the framework of information theory, conceptualizing brain cells as hubs for processing information through nervous signals and stimulations. In the realm of this burgeoning information theory, Sherrington proposed three mechanisms for integrating information within the central nervous system's cell network. Studying basic neural circuits linked to reflex responses, Sherrington sought to comprehend how information propagated between synaptic terminations and subsequent nerve cells within the interstitial space termed as synapses. He hypothesized that although every electrical action in the brain should generate a motor output, not all nerve activity was excitatory; it also involved inhibitory inputs. Each neuron receiving nerve signals would need to assess the relative weight of excitatory and inhibitory afferents, engaging in integration activities before making a "decision" to initiate or suspend a subsequent electrical impulse. Subsequent studies by Eccles (Bradley &

Eccles, 1953), employing intracellular recordings of ionic mechanisms, validated this hypothesis: the integration of excitatory and inhibitory signals results in the action potential discharge only if excitatory afferents outweigh inhibitory ones.

The inquiry into the nature of neuron communication, however, persisted. In the eighteenth century, Luigi Galvani detailed how the muscles of a frog's amputated leg, suspended from an iron lattice with a brass hook, could contract during a thunderstorm with lightning. Galvani later demonstrated that he could induce the same contraction experimentally, effectively inventing the first electric battery in history. In the nineteenth century, Carlo Matteucci created a form of bio-battery using a series of muscle tissues from frog thighs connected to form a muscle pile. He subsequently replicated the experiment with eels, pigeons, and rabbits. Although the interpretation of these phenomena remained distant from the scientific mechanism of the action potential and leaned toward metaphysical explanations, these experiments laid the foundation for von Helmholtz's investigations. Von Helmholtz measured the rate of electrical transmission in frog muscle fibers and nerves of varying lengths. Despite an impressive speed of approximately 40 m/s, he observed that this speed was lower than the propagation speed of electricity. Consequently, the transmission of information in nerve cells had to occur differently from electrical propagation, involving electrochemical reactions that actually took longer than passive electrical conduction.

In the 1940s and 1950s, Hodgkin, Huxley, and Katz elucidated the mechanism of action potential propagation along the axon without loss of potential. This process occurred unidirectionally, mirroring the flow of ionic current. The ionic hypothesis posited that the resting membrane potential resulted from passive potassium-permeable channels. Simultaneously, the action potential originated from two distinct conduction pathways: one selective for sodium and the other for potassium. The ionic hypothesis spurred studies that enabled the description of membrane proteins and their mechanisms of activation and inactivation, ultimately leading Fatt and Katz (1951) to propose the initial formulation of the theory of synaptic transmission. According to this theory, the transmission of information between neurons is facilitated not only by electrical sodium-potassium channels but also by proteins capable of ligand-dependent activation. In other words, these proteins respond to chemical stimulations by specific molecules. Eccles (1964) later extended Fatt and Katz's early observations of peripheral synapses to all other synapses in the nervous system. In addition, Changeux (Changeux et al., 1992) described how in ligand-dependent channels, the neurotransmitter binding site and ion channel constitute different domains within the same membrane protein.

Hence, inter-neuronal communication involves the discharge of chemicals from sites located in the presynaptic terminal of the axon. When the action potential reaches these sites, these chemicals, known as neurotransmitters, are released into the extracellular space between synapses. In the 1920s, Otto Loewi conducted an experiment involving two frog hearts, where the nerves were connected to only one of the hearts. Placing both organs in saline solutions, he electrically stimulated the nerves of the first heart, initiating its heartbeat through efferent fibers. After removing the saline solution from the stimulated heart, Loewi injected it into the second heart, which, too, exhibited a heartbeat as if stimulated by an electric current. This observation suggested the release of some chemical into the saline solution. Thus, the activation and release of neurotransmitters by vesicles enable the initiation of an action potential in the postsynaptic cell, facilitating the propagation of information between cells in the nervous system.

Chemical brothers

Excitatory nerve cells rely on specific messengers, known as neurotransmitters, to transmit excitatory neural messages to the postsynaptic cell. The primary excitatory messenger in the central nervous system is glutamate, an amino acid neurotransmitter. Glutamate facilitates the rapid activation of action potentials, exerting excitatory effects and participating in metabolic regulation processes. In contrast, the principal inhibitory neurotransmitter is gamma-aminobutyric acid (GABA). When GABA is released into the inter-synaptic space, it decreases the likelihood of postsynaptic cell activation, effectively reducing the possibility of initiating an action potential.

The actions of both glutamate and GABA are integral to neurotransmitter functioning in the central nervous system and depend on their binding affinity with specific postsynaptic receptors. Various receptors are present for different neurotransmitters. Glutamate binding to postsynaptic receptors leads to the closure of a synaptic passage, causing the movement of positive ions from the extracellular fluid into the cell. This alters the chemical balance between the inside and outside of the postsynaptic cell, resulting in a positive internal voltage sufficient to initiate an action potential, typically around −60 mV at rest. Conversely, GABA receptor occupation induces depolarization, decreasing the chances of triggering postsynaptic potential.

Beyond glutamate and GABA, other neurotransmitters modulate their excitatory and inhibitory activity. Peptide-type neurotransmitters, such as endorphins and enkephalins, and monoaminergic neurotransmitters, such as norepinephrine, adrenaline, dopamine, and serotonin, as well as cholinergic neurotransmitters (e.g., acetylcholine) and amino acid neurotransmitters (e.g., aspartate and glycine), play roles in this modulation. Peptides, released in response to experiences of pain and stress, can either enhance or diminish the effects of GABA and glutamate. Monoamines, produced in the brainstem with axon projections to various brain areas, modulate the effects of GABA and glutamate, influencing physiological activation or arousal states in the body. Neurotransmitters such as adrenaline, norepinephrine, dopamine, and serotonin are widely involved in behavioral and socio-emotional regulation during stressful situations and mood states.

Hormones, specifically cortisol and oxytocin, constitute another class of neuromodulators with relevance in developmental psychobiology. Cortisol, known as the stress hormone, is a steroid hormone released in response to stress. It affects the central nervous system and mobilizes energy needed for motor and behavioral responses to cope with stressful situations. Oxytocin, often termed the attachment hormone, is associated with affective bonding, particularly active in the maternal brain postnatally. Intranasal administration of oxytocin can induce subjective sensations of socio-affective affiliation.

The assembly

Today, our understanding of the nerve cell, or neuron, reveals a dual composition: the central part is known as the soma and the peripheral components are termed nerve fibers. The soma handles crucial managerial functions, including genetic material storage, protein synthesis, and cell survival regulation, resembling the behavior of other cells. In contrast, nerve fibers, comprising axons and dendrites, facilitate the coordinated functioning of neural cell groups. Axons, acting as efferent channels, terminate with synaptic connections, allowing communication with recipient neurons. Dendrites, serving as afferent channels, possess spines crucial for receiving inputs and contributing significantly to brain development, learning, and memory.

The principle of divergence dictates that a neuron has one axon that can bifurcate into multiple terminals, while the principle of convergence requires each neuron to receive more afferents from other nerve cells. Circuits consist of neurons connected by synaptic connections, forming systems serving specific functions like sight or hearing. The interaction between interneurons and projection neurons within these clusters is vital. Projection neurons, with long axons extending beyond their cell bodies' localization, release neurotransmitters to stimulate the next cell in the projection hierarchy. Interneurons, responsible for short-range communications, facilitate information processing within a circuit and regulate projection neuron activity.

Our brain, comprising approximately 100 billion nerve cells, forms connections with an average of 10,000 neurons per cell. Brain circuits develop and change in response to experiences, influencing neural connections and organization. Hebb's principle underscores that repeated experiences shape our brain's connections over time. Lack of stimulation and experience can lead to cell death through synaptic pruning, determining the expression of genes and ultimately shaping the brain's architecture. Life experiences significantly impact brain development, especially sensory and perceptual systems and specialized circuits sensitive to learning through use.

Although genetically determined pathways dictate brain structure differentiation, early experiences play a crucial role, particularly in sensory and perceptual system maturation. Wiesel and Hubel's studies illustrate how a lack of specific early visual experiences alters visual functioning and neural architecture. Plastic adaptation in the central nervous system results in a general reorganization of connections, reflecting principles of non-linear dynamical systems.

Considering the brain as a dynamic system within the broader system of the human being, it operates as an integrated, complex system shaped by continuous interactions between components influenced by experience. Genes, regulating neural development and function, contribute to the system's operation. Experiences influence how genetic instructions are read, interpreted, and translated into observable phenotypes over time through epigenetic processes. These processes, governing genetics "from the outside," yield diverse phenotypes from a single genotype, emphasizing the role of experiential learning in achieving adaptation, complexity, and system coherence.

Learn like a slug

How is it conceivable that experiential learning wields such influence over brain development, self-evolution, and the discernible behavioral traits of an individual? Learning, essentially, constitutes a modification of behavior or a parameter adjustment within the system resulting from the assimilation of fresh insights about the world. According to Kandel, the ability to learn from experiences stands out as the most remarkable facet of human behavior, playing a significant role in shaping who we are. Kandel, a Nobel Prize recipient in Medicine in 2000, has extensively studied two intriguing forms of learning. The first, habituation, represents the most straightforward type of learning. It involves the diminishing of a specific behavioral response upon repeated exposure to a stimulus. Initially, a new stimulus grabs our attention, elicits surprise, fear, activates us, or directs our focus, leading to physiological changes like increased heart rate and accelerated breathing. We become alert. However, with repeated exposure to the same stimulus, we swiftly learn to recognize, label, and reduce our bodily and physiological reactivity. Without this capacity for habituation, navigating the environment, such as living in a bustling city like Milan, would be extraordinarily challenging for any living being. Habituation, therefore, entails the ability to identify and dismiss stimuli that are no longer novel or significant. The study of habituation, as we understand it today, started in the '60s with observations on the response to subsequent stimulation of motor neurons through intracellular recordings. The inhibition of response behavior, exemplified by the stretching reflex in cats, resulted from a reduction in synaptic convergence to motor neurons.

Nevertheless, cats present a neurological complexity that makes isolating a single circuit challenging. Consequently, Kandel turned his attention to studying habituation and other forms of short- and long-term learning processes in a simpler animal model—the sea snail, *Aplysia californica*. This creature exhibits a gill retraction reflex, similar to the defensive reflexes found in mammals. Previous studies had already demonstrated how this reflex could be regulated by habituation processes (Kandel, 1976). The underlying neural circuit comprises merely 6 motor neurons, 24 sensory neurons directly connected to motor neurons, and additional excitatory interneurons. Many of these cells are substantial and easily distinguishable (Frazier et al., 1967). The application of a tactile stimulus to stimulate sensory neurons triggers the release of a chemical transmitter, which interacts with the receptors of the motor neuron, reducing its membrane potential and initiating the action potential. Kandel observed that with the initial stimulation of the sensory neuron, a significant excitatory synaptic potential occurred in motor neurons, resulting in rapid gill retraction. However, through repetition, the number of action potentials diminished, and the response became less frequent until the reflex behavior vanished entirely. Notably, after only 10 stimulations, there was an initial occurrence of short-term habituation. This stemmed from the fact that the excitatory chemical synapses between sensory neurons and motor neurons lost their functional effectiveness with repeated stimulation, releasing a smaller quantity of neurotransmitters. Furthermore, Kandel and colleagues demonstrated the possibility of inducing long-term habituation. Before habituation, 90% of sensory neurons formed individual connections on motor neurons. However, after long-term learning, only 30% of sensory neurons established connections with motor neurons, and even after three weeks, these connections were only partially restored.

Another type of learning is sensitization, which stands in contrast to habituation. In sensitization, an organism learns to intensify its behavioral response following a harmful or unexpected stimulus. In a sense, sensitization can be viewed as a form of learned fear, where the organism adapts to react more strongly to a specific stimulus that would typically be neutral under different circumstances (Pinsker et al., 1970). Sensitization involves the activation of a mechanism termed as presynaptic facilitation by Kandel. In this process, neurons that mediate sensitization converge on the endings of sensory nerve cells, enhancing their ability to release neurotransmitters. Consequently, the synaptic terminal undergoes regulation in the opposite direction in comparison to habituation. In presynaptic facilitation, the neurotransmitter, primarily serotonin, acts on the endings of sensory neurons, elevating cyclic AMP levels. This, in turn, facilitates membrane channel phosphorylation and increases calcium influx, resulting in a higher release of neurotransmitters (Klein & Kandel, 1978). Throughout the sensitization process, the stimulus inducing anxiety in *Aplysia* promotes the connections established by sensory neurons on their target cells, namely interneurons and motor neurons.

Another type of learning investigated by Kandel in the sea snail involves classical conditioning. The distinguishing factor from sensitization is the introduction of a conditioned stimulus or signal, specifically associated and administered in a contingent manner with the aversive stimulus. Similar to humans, *A. californica* exhibited behaviors such as escape movement, resembling the anticipatory anxiety response, or learned fear in reaction to adverse stimuli, such as a strong shock to the head. Notably, given that the sea snail is an herbivore feeding exclusively on algae, Kandel identified a neutral chemical stimulus in shrimp extract that could serve as a conditioned signal for learning. The sea snail demonstrated the ability to learn the conditioned response: subsequent presentations of the signal, even in the absence of the aversive stimulus, triggered the motor escape response, indicating the establishment of an anticipatory anxiety or learned fear response.

Beyond learning: Recollection

Kandel's studies provide a framework for observing the Hebbian principle of neural learning (Hebb, 1949), a well-established concept that describes the regulation of neuroplasticity in our brain. The term "plasticity" was likely first introduced by

Konorski, a Polish neuroscientist, to characterize the ability of nerve cells to be altered by experience. However, the principles of plasticity and learning fall short of capturing our ability to derive value from experience: we learn from our interactions with the environment not only by modifying the functioning of our brain but, more importantly, by ensuring that these modifications become relatively stable and integrated into the overall organization of our brain. It seems inappropriate to label a modification or change that doesn't endure over time as true learning; perhaps, it could be better described as a contextual response.

As is often the case in scientific progress, a crucial advancement in our comprehension of complex processes can be attributed to a serendipitous observation. Terje Lømo, a Norwegian PhD student in physiology conducting research at Andersen's laboratory in Oslo in 1966, made a noteworthy discovery. While investigating the effects of electrical stimulation on afferent fibers of the hippocampus in the brains of anesthetized rabbits, Lømo noticed that short sequences of closely spaced stimulations led to an increased efficiency of nerve transmission that could endure for several hours (Lømo, 2003). It is essential to note that during those years, there was already a hypothesis suggesting the hippocampus's potential role in memory, both in animal models and humans. Despite Lømo's observations not immediately garnering attention in the scientific community, and the Norwegian physiologist himself not publishing them until later, an English researcher named Bliss, a student of Hebb at McGill University in Montreal, joined the same laboratory as Andersen a few years later. Collaborating with Lømo, Bliss expanded on the initial observations, and together, they published the first study defining what is now known as long-term potentiation (LTP) in 1973. They validated Lømo's early findings and demonstrated how a brief burst of high-frequency pulses could serve as an enhancing stimulus to facilitate synaptic connections. It soon became evident that this mechanism could be a universal process of neural learning and memorization, where the central nervous system's activity in response to external stimuli would lead to enduring alterations in synaptic associative function. In the subsequent years, this discovery sparked a surge of studies and publications on the subject, further facilitated by subsequent researchers confirming the feasibility of studying this mechanism not only in live animals under sedation but also in sections of hippocampal tissue.

The specific nature of synaptic facilitation following LTP was thus confirmed. In fact—and this represents a confirmation and corollary of Hebb's axiom—the potentiation concerned only the afferent neural lines that were experimentally stimulated, while other afferents to the postsynaptic cell were not affected. These studies further confirmed the role of the environment in the education—not only in the selection—of the organization and functioning of the central nervous system. A further specification of the Hebbian learning principle that we owe to LTP studies concerns the possibility that this enhancement occurs specifically if two nerve pathways are stimulated simultaneously and contingently. This principle of co-operation represented a confirmation of Hebb's axiom that contingently stimulated cells develop a long-lasting association that involves the activation of one in conjunction with future activations of the other. The connection, as anticipated in

Chapter 2, is, in fact, the secret of our organization as complex systems, and the same can be said for our brain, a dynamic, complex, and non-linear tool refined by evolution.

What are the changes that occur at the synaptic level and that are responsible for strengthening? First, this process must have to do with the functioning of receptors in the postsynaptic terminal, since manipulation of glutamate receptors in the target cell—although it does not interfere with the proper functioning of nerve transmission—prevents the facilitation of cellular association and co-operation. In addition, Lynch and colleagues (1983) further demonstrated that it was some mechanism at the level of channel protein functioning that facilitated the LTP process since the impediment of calcium influx into the postsynaptic cell seemed to limit the establishment of LTP. Studies of LTP in glutamate-containing cells have shown that different types of receptors are implicated in experience-dependent enhancement of synaptic connections. It is the differential affinity of glutamate with two types of receptors (NMDA and AMPA) that favors potentiation: the AMPA receptor receives the signal and helps to initiate the action potential, while the NMDA receptor favors the influx of calcium, a necessary condition for LTP and Hebbian associativity. NMDA receptors are activated only through stimuli that we can define as strong—for example, the high-frequency stimulation train of Lømo and Bliss—and act as "contingency detectors" or coincidences: they record the simultaneous and co-operative activity of presynaptic and postsynaptic neurons at the time of neural discharge.

Subsequent studies on this type of molecular mechanism provided further support for the hypothesis that LTP could be involved in the cellular functioning of short- and long-term memory. In the 1980s, studies on animal models had shown that the blockade of protein synthesis could negatively affect the ability to form long-term learning but not short-term behavioral changes, suggesting that the two mnemonic processes could have a different biochemical substrate. Kandel and colleagues have shown how a genetic alteration of specific kinases—protein kinase A and calcium-calmodulin-dependent protein kinase—can induce long-term impairment of LTP, while short-term memory would be preserved (Mayford et al., 1995).

Short-term memory would be allowed by the action of specific enzymes—protein kinases—which have the task of phosphorylating proteins, that is, activating them. When the influx of calcium that occurs during LTP activates protein kinases, it is, in particular, the AMPA receptor that is phosphorylated, and from that moment, the same amount of neurotransmitter can more easily release an action potential in the postsynaptic cell. Long-term memory, on the other hand, would involve not so much the phosphorylation of proteins in the postsynaptic cell but the synthesis of new proteins, called kinases. These would be able to activate, in the cell nucleus, a cell transcription factor called CREB (cAMP response element-binding protein) (Kandel, 2012), which, in turn, would stimulate the gene transcription of specific new proteins that act on the terminal of the postsynaptic cell, stabilizing the synaptic connection.

In the "return" journey from the cell nucleus to the dendritic terminal, these proteins can recognize the terminal of interest—the one involved in the connection to be enhanced—through specific molecular markings (Kandel, 2012) and favor the enhancement through two mechanisms. First, they make more AMPA receptors available on the receiving surface, so that the same amount of neurotransmitter is associated with a facilitation of synaptic transmission. In addition, the LTP process could also promote synaptogenesis, the creation of new connections through the release of neurotrophins at the level of dendrites. In addition, the role of NMDA receptors—as anticipated—is documented in the development of our memory capacity. Mice treated with substances that inhibit and block NMDA receptor activity in the hippocampus show difficulty in making spatial learning within water mazes in experimental environments (Shinohara & Hata, 2014). These results confirm previous results of studies involving mouse genetic models in which building blocks of NMDA receptors were subtracted (Ebralidze et al., 1996).

Other studies have shown that the same mechanisms that regulate the formation of short- and long-term memory in the hippocampus can also be applied to the availability of emotional memories involving other areas of the brain. In a series of experiments, Rogan and colleagues (1999) showed that LTP could also occur in the amygdala in response to aversive stimulations accompanied by a sound stimulus. In addition, fear conditioning—through the association between auditory stimuli and electrical stimulation—produced LTP in the cells of the amygdala responsive to acoustic stimuli. Certainly, the circuits of the hippocampus and amygdala are very different and have different connections with the sensory nuclei, with the cortex, and the brainstem. In addition, several neurotransmitters—such as GABA (Lucas & Clem, 2018) and serotonin (Gebhardt et al., 2019)—have recently been associated with memory formation in fear conditioning. However, the calcium-permeable nature of NMDA receptors appears to be a ubiquitous feature within the central nervous system (Alkadhi, 2021). Consistently, it is highly probable that the same molecular mechanisms, therefore, facilitate short- and long-term learning in different brain areas and not only for spatial but also emotional forms of memory (LeDoux, 2000).

An integrated systemic view of developmental psychobiology

This chapter marks the culmination of the foundational section of the book, dedicated to unraveling the fundamental principles essential for comprehending how diverse biological, neurophysiological, and endocrinological systems collaboratively contribute to shaping a child's early development and determining their developmental trajectories. By conceptualizing the child as a dynamic system capable of non-linear adaptations within the context of life, we gain a nuanced understanding of the role of psychobiological regulations throughout the developmental journey, steering clear of potential oversimplifications. The child's receptiveness to environmental stimuli, influenced by both phylogenetic and evolutionary adaptations (as seen in Taung's baby) and ontogenetic predispositions (as observed in the proficient fetus), serves as the groundwork for psychobiological regulations to impact systemic oscillations, instigating phase changes and phenotypic variations.

As we transition to the next section, experiential learning trajectories will continue to shape selection and information processes within the central nervous system. The ongoing regulation of interconnected parameters will facilitate the emergence of adaptive solutions, not easily foreseeable, allowing the organism to integrate proprioceptive and exteroceptive signals into a cohesive whole—forming what is commonly referred to as the self. This lays the foundation for delving into the dynamic processes explored in the upcoming section on developmental psychobiology.

Section II

Processes

4 Social by evolution

In 2013, Thomas Insel, then director of the National Institute of Mental Health (NIMH), made a significant announcement. He declared that NIMH would no longer fund research projects based solely on the Diagnostic and Statistical Manual of Mental Disorders (DSM) logic, as he found the diagnostic categories in the DSM to be invalid and inappropriate. Insel emphasized that "patients deserve better" (McCarthy, 2013). This shift in perspective criticized the symptom-based diagnostic system and paved the way for a vision that sought to understand the pathology of patients through solid genetic and neuropharmacological data. However, this institutional turn faced resistance from psychotherapists, especially those from the psychodynamic school. They labeled this approach as reductionist, arguing against the idea that all clinical manifestations should be solely attributed to cerebral or genetic markers. In this context, Stephen Porges published an updated Polyvagal Theory in 2011, which attempted to place relationships back at the center of scientific and clinical understanding of mental disorders while also exploring the deep connection between evolution and social development at the neurophysiological level. Porges had initially introduced his ideas in the early nineties through an article titled "Orienting in a Defensive World: A Polyvagal Theory," published in *Psychophysiology* in 1995. Together with Sue Carter, Stephen Porges has become one of the most influential thinkers and researchers in the field of developmental psychobiology, significantly impacting scientific knowledge and psychotherapy practices.

This chapter aims to introduce the evolutionary assumptions of polyvagal theory and its implications for developmental psychobiology. However, Porges' theory has not been exempted from criticism, particularly regarding the evolutionary basis of his ideas, which can be challenging to falsify, as discussed in the previous chapter. Other critical points revolve around the quantitative indices used as an indirect measure of the theory's assumptions and the consistency of empirical evidence. Therefore, the chapter will also provide a brief review of the criticisms directed at this theory. Finally, the chapter will explore some of the clinical implications of polyvagal theory.

The Da Vinci code

Between 1504 and 1506, Leonardo da Vinci created one of the earliest hand-drawn depictions of the vagus nerve, which is now known as the reversible nerve. In his

DOI: 10.4324/9781003479314-6

notes, the Italian artist and scholar argued that if the motion of the heart originated from the "reversive" nerves in the brain, then it would become clear how the soul, or animal spirits as defined at the time, had its origin in the left ventricle of the heart. Darwin (1872) had already hypothesized about bidirectional communication between the central nervous system and the heart, suggesting that the dynamic relationship between the vagus nerve and the central nervous system played a role in the spontaneous expression of emotions. According to Darwin, when an emotional state is established, the heartbeat changes instantaneously and influences brain activity, particularly in brainstem structures via the vagus nerve. While Darwin did not elaborate on the specific mechanisms involved, his proposal emphasized the regulatory role of the pneumogastric nerve, which was later renamed the vagus nerve toward the end of the nineteenth century.

Darwin's proposal was also advocated by his contemporary, the French physiologist Claude Bernard. Often regarded as the "real father" of the concept of homeostasis, Bernard initially defined it as "milieu intérieur," or "half internal" in French. With this concept, Bernard suggested that if the organism's functioning depends on physical conditions and chemical regulations at the interface between the environment and the interior of the organism itself, then these conditions and regulations should remain as constant as possible. Bernard acknowledged the dynamic nature of homeostasis, proposing that maintaining a constant internal environment required continuous adaptations by the organism to compensate for environmental fluctuations (Bernard, 1878–1879). The concept of homeostasis was further refined by Cannon in 1929, who described the dynamic regulatory processes that allowed the organism to achieve the constant internal conditions proposed by Bernard. Bernard himself (1865) considered the heart as a primary response system capable of coping with various forms of sensory stimulation.

In some ways, Darwin and Bernard can be considered the founding fathers of modern psychophysiology, as they contributed to the idea that the regulation of cardiac activity not only results from brain activity but it also serves as a source of afferents to the central nervous system. This concept is reflected in later proposals, as well as in Porges' polyvagal theory (1995). During the 1960s, physiological scholars had limited theoretical constructs and tools at their disposal. A vague theory of physiological arousal, encompassing sensory and visceral activation, was dominant, but there was no clear understanding of the anatomical, physiological, and functional components of arousal. It was generally assumed that this physiological activation was solely mediated by the sympathetic nervous system. One of the founders of psychophysiology proposed considering a continuity between cortical activation measured by electroencephalogram and peripheral physiological activation, often measured by skin conductance at the level of the hands. This view of a peripheral indicator of brain activity aligned with Pavlov's proposal for measuring the autonomic system in classical conditioning experiments, where behavioral responses were regarded as observable manifestations of underlying brain processes. During this time, physiological measures were considered a suitable approach to indirectly investigate psychological processes that did not require explicit awareness or verbal responses, such as emotions.

In the context of nascent psychophysiology as an interdisciplinary discipline encompassing psychology, medicine, physiology, and engineering, the first issue of the scientific journal *Psychophysiology* was published in 1964. This journal primarily focused on studies where psychological variables were treated as independent variables to examine the physiological outputs or outcomes of mental processes (Porges, 2011). This approach differed from psychological physiology, which aimed to manipulate physiology and observe its consequences on the behavioral level, specifically arousal. Despite the continued existence of a dualistic perspective between mind and body, physiology and psychology, the emergence of other interdisciplinary fields such as cognitive and socio-affective neuroscience or developmental psychobiology suggests that this distinction may only be useful for educational purposes. Stephen Porges' polyvagal theory challenges this dualism, offering a model of bidirectional interaction between the brain and physiology. It interprets the brain's contribution to peripheral regulations as an evolutionary platform enabling the emergence of human-specific social behaviors and promoting adaptation.

The vagal paradox

The roots of polyvagal theory date back to 1969 when Porges published his master's thesis with Raskin, focusing on the role of interbeat heart rate in attentional states. However, it was only toward the end of the 1980s that Porges could provide a reliable, albeit indirect, measure of the activity of the parasympathetic system. His theory was fully formalized in the mid-1990s (Porges, 1995). A pivotal moment in Porges' theorization occurred in 1992 when he received a letter from a neonatologist criticizing the idea that measuring vagal tone from interbeat heart rate could serve as an index of the clinical well-being of newborns. Porges had reported high levels of vagal tone in healthy infants but low levels in at-risk infants born preterm. However, the neonatologist provided clinical evidence that high vagal tone values were actually a risk factor for bradycardia conditions that could lead to infant mortality. This vagal paradox led Porges to delve into extensive research, devouring hundreds of articles and books on the neural regulation of the autonomic nervous system in vertebrates at the library of the National Library of MNIH during his visiting period.

To resolve the vagal paradox, Porges hypothesized that in mammals, there are two different anatomically distinct vagal response systems. This hypothesis is based on the fact that the vagus nerve is not unitary; rather, it consists of a family of neural pathways originating in different areas of the brainstem. Moreover, the vagus is not solely an efferent pathway; approximately 80% of vagal fibers are afferent, carrying information from the periphery to the central nervous system (Agostoni et al., 1957). In addition, the vagus is lateralized, with two parallel bundles, right and left, which are asymmetrical. The right vagus has a greater influence on the chronotropic regulation of the heart, i.e., the heart rhythm. Essentially, Porges argues that mammals are polyvagal (Porges, 2011). During the orientation reflex, there would be a reduction in vagal tone in one branch of the vagus, leading to bradycardia, while another branch would facilitate attention. Porges supports this position using anatomical data derived from the animal models of reptiles and mammals.

Of reptiles and mammals

In mammals, the vagus nerve's principal motor fibers have their origins in two separate nuclei: the dorsal nucleus and the ventral nucleus, also referred to as the ambiguous nucleus. A third nucleus, the solitary tract, receives numerous afferents from the body's periphery and establishes connections with the vagus nerve. The dorsal nucleus primarily sends projections to structures located below the diaphragm, such as the stomach and intestines, while the ambiguous nucleus predominantly innervates structures above the diaphragm, including the larynx, pharynx, palate, esophagus, bronchi, and heart. Although both nuclei receive afferents from the solitary tract, amygdala, and hypothalamus, they appear not to have direct connections with each other. In reptiles, the distinction between the dorsal and ambiguous nuclei is not as clear, and cardiac vagal efferents mainly originate from the dorsal nucleus. In mammals, however, there is a distinct anatomical difference between the two nuclei, with the ambiguous nucleus serving as the primary source of vagal efferents to the heart. In addition, the ambiguous nucleus has direct connections to the amygdala and specific efferents to the facial muscles (Porges, 2011).

Functionally, Porges presents evidence to support the different regulation of the vagal system in reptiles and mammals. Reptiles use projections from the dorsal nucleus to the heart for specific events that require motor activation, such as orientation, immobilization in response to a predator, or conserving oxygen for long periods. In contrast, mammals use the vagal efferents of the ambiguous nucleus to maintain a constant brake on metabolic potential inhibition. This higher vagal tone is observed during safe situations, like sleep, while it is reduced in response to environmental demands that require metabolic activation. Unlike reptiles, mammals have more motor and behavioral repertoires to respond and orient themselves to unexpected or new environmental stimuli. They can voluntarily maintain attention, seek more information, and express regulation through facial and verbal expressions. In addition, mammals have a higher energy load and metabolic demands compared to reptiles. For instance, mammals with a heart divided into four chambers, like humans, can continue significant motor activities even during digestion.

Reptiles have reduced vagal tone under basic environmental conditions, transitioning to increased vagal tone in response to stimuli or threats. In mammals, vagal tone is elevated during restful conditions and reduced in response to environmental demands. In reptiles, social interactions like parenting or reproduction are minimal, making bradycardia an adaptive response that doesn't compromise their physiological state. On the other hand, mammals must increase metabolic output to activate adequate responses to the environment. Sustained bradycardia in response to unexpected stimuli would reduce oxygen supply and metabolic efficiency, compromising the mammalian body's survival chances. Therefore, mammals require a physiological response that doesn't compromise oxygen requirements, achieved by the removal of the vagal brake originating in the ambiguous nucleus, as proposed by Porges.

Polyvagal

Porges' theory proposes that through evolution, mammals have developed two vagal systems: a phylogenetically ancient vagal system inherited from amphibians and reptiles, coupled with a more recent vagal system typical of mammals from an evolutionary point of view. In many vertebrates, the primary response systems to a hazard are the fight-or-flight reaction and immobilization. The first reaction allows the body to defend itself or escape a perceived threat. These behaviors require rapid access to resource mobilization through metabolically costly activation of the sympathetic nervous system. Immobilization, or freezing, is instead an even more ancestral system shared with all vertebrates. Unlike the more expensive fight-or-flight strategies, immobilization is an attempt to reduce metabolic demands for food or oxygen and to appear apparently inanimate, pretending to be dead. It is a strategy that requires a massive reduction in autonomous function through vagal activity of the parasympathetic nervous system.

Over the course of evolution, a second vagal system evolved and provided the ability to control and regulate both defense strategies. This second component of the parasympathetic system is observed mostly in mammals. The anatomical structures that regulate this component of the vagus are located in the brainstem and provide innervations that reach the muscles of the face and head, creating an integrated system of social involvement. The outputs of this system consist of motor pathways that regulate the voluntary striated muscles and the heart muscle. With phylogenetic evolution, the organization of the vagal system has become increasingly complex, incorporating nerve projections that include the trigeminal, facial, and glossopharyngeal muscles. Specialized functions, such as the rotation of the head in response to sensory stimuli to identify the source of stimulation, chewing, and salivation to facilitate the gustatory and digestive processes, are further integrated into the vagal system. These ascending fibers that mostly innervate the facial muscles are generally referred to as special visceral efferences to distinguish them from the general visceral efferences that innervate the heart muscle. These cranial nerves always originate in the ambiguous nucleus, which in turn receives an important source of information from the trigeminal nerve. Control of muscles above the diaphragm appears to be typical of mammals and coordinates complex sequences of sucking, chewing, digestion, and breathing, as well as facial expressiveness in social settings.

Central to Porges' theoretical proposal and its validation is the availability of a reliable measure of vagal tone. Porges argues that the ability of the ambiguous nucleus to regulate both general and special visceral efferences can be monitored and quantified in terms of the amplitude of sinus respiratory arrhythmia (RSA). The vagal fibers that originate in the ambiguous nucleus and have an inhibitory action on the sinus-atrial node reflect the trend of the respiratory rhythm and therefore produce RSA, which stands for respiratory sinus arrhythmia. RSA, would therefore, be a measure of the general visceral efferences of the ambiguous nucleus that seems to correlate almost completely with respiratory rhythm measures (Porges & Bohrer, 1990). The suppression of vagal tone reflects a removal of the vagal brake, which is the tonic inhibitory influence of the myelin vagal on the heart, slowing the intrinsic rhythm of the cardiac pacemaker. When vagal tone decreases its influence on the heart—the vagal brake is released—the heart rhythm increases spontaneously. It is not due to sympathetic arousal, but to the release of the vagal brake that allows the intrinsic heart rhythm to express itself.

Distinguishing foes from enemies

Porges believes that by processing information received from the environment through the senses, the nervous system is engaged in continuous risk assessment. He refers this process as "neuroception" to describe how neural circuits distinguish situations or people with whom we can feel safe from contexts in which we perceive elements of danger or threat to our integrity. Being a process inherited by our species through the vicissitudes of evolution, neuroception would take place in primitive parts of the brain, outside of our awareness. Even when we do not yet realize it, our body would then start processing sensory information to direct our response at the visceral level in the direction of an attack, escape, or freezing. This process would explain why a child vocalizes in an affiliative way toward a familiar adult but cries as soon as a stranger approaches or why an eighteen-month-old baby gets involved with pleasure in the parent's embrace but interprets the same gesture as threatening if carried out by an unknown adult. Prosocial behavior would therefore not occur—according to Porges—without the intervention of a neuroception process that conveys the feeling of security and supports the inhibition of defensive strategies. In order to allow the body to implicitly select defensive strategies or affiliation behaviors, however, it is necessary for the nervous system to do two things: assess the risks and inhibit the primitive responses of attack, flight, or freezing.

   A process of fallacious neuroception—that is, inaccurate in assessing the safety or dangerousness of a context—could contribute to the emergence of maladaptive ways of regulating social exchanges and the expression of defensive behaviors associated with specific psychiatric disorders. Porges cites as an example the inhibition of contingent responses of attack, flight, or immobilization in children with autistic syndrome and reports the results of studies conducted on institutionalized children of the Bucharest project, where children with reactive attachment disorders tend to be strongly inhibited or uninhibited with respect to social involvement (Zeanah, 2000). Particularly, in one of these studies, a group of children raised in orphanages were cared for by 20 caregivers who rotated in shifts, resulting in a ratio of about 3 caregivers to 30 children in each shift. A second group consisted of a smaller turnover, with 10 children cared for by 4 caregivers. According to Porges' theory, it should be expected that familiarity with caregivers should promote children's sense of security, facilitating the appropriateness of social and affiliative development. The results of Smyke and colleagues (2002) support this hypothesis:

the higher the frequency of contact with specific caregivers, the lower the incidence of reactive attachment disorders in this group of institutionalized children.

When our system picks up safety elements in the environment, it regulates metabolism accordingly: heart rhythm decelerates and stress response systems (see Chapter 5) are turned off. The process of neuroception would be possible through the recruitment of specific areas of the brain that can process body movements, facial expressions, and vocalizations of another person, thus contributing to developing a perception below the threshold of awareness of safety and trust or danger and threat. Porges believes that if our ability to engage in social relationships were to depend solely on a voluntary motor basis, then the newborn would have a number of disadvantages in initiating their participation in interactive exchanges. In fact, conscious control of motor components is immature at birth and takes several months to develop satisfactorily. In Porges' model, access to social skills does not depend on how well we modulate and regulate voluntary musculature, but on the innervation of facial muscles that are intrinsically linked to the cortex and brainstem. The neural pathways that connect the cortex to these nerves in the trunk are sufficiently myelinated at birth and allow the baby to send signals to the parent by vocalizing or through facial expressions and to engage in social connections through looking, smiling, and sucking.

From this point of view, it is important to remember that if the neuroception process hypothesized by Porges was really in operation from birth, it could contribute to understanding the medium- and long-term effects observed in socio-emotional development in children exposed to severely depressed parents. In these situations, although there are no conditions of serious danger or threat to survival, the neuroception of the child would lack those signs of social contingency and affective reflection that are necessary to identify the situation as protective and safe, and therefore the behaviors of affiliation and social involvement would be inhibited. This hypothesis is indirectly supported by evidence that children of depressed mothers show difficulty developing dyadic coordination states during brief face-to-face interactions, as in the case of the still-face paradigm (Tronick & Weinberg, 1997).

Critiques

As will be reported later in this chapter, Porges' theory has been a great success on the scientific level and for its clinical implications. However, it has not been exempted from criticism, mostly due to the difficult falsifiability of its evolutionary premises. Paul Grossman (Grossman & Taylor, 2007) is perhaps the greatest critic of Porges' theoretical proposal and still maintains a page on one of the main social networks for researchers—ResearchGate—in which he reports on the evidence that undermines the rigor of the polyvagal theory premises.

Grossman criticizes the assumption that the anatomical distinction between the dorsal and ambiguous nuclei of the brainstem is only present in mammals (Porges, 2003), as well as the implicit premise that RSA is always an index of the vagal tonic activity that originates in the ambiguous nucleus. Grossman, in particular, provides evidence of the presence of a dual localization of vagal neurons in other vertebrates, such as amphibians and turtles. For example, the Mexican ambystoma—a specimen of neotenic salamander generally called axolotl—would show a rudimentary nucleus of vagal cells attributable to the ambiguous nucleus but located at the level of the dorsal nucleus (Barriga-Vallejo et al., 2015). In fish and crocodiles, the percentage of neurons that can be located in the ambiguous nucleus is around 10%; it is lower in lizards and birds. It would, therefore, be possible to trace a dual topography—dorsal nucleus and ambiguous nucleus—in amphibians and reptiles also, not only in humans or mammals in general.

There has been contention regarding whether the intermittent control of heart rhythm, leading to the respiratory oscillation reflected by the RSA index, is exclusive to mammals. The idea of a similar connection between intermittent regulation and ventilation, known as cardiorespiratory synchrony, has been historically proposed for fish (Satchell, 1960) and certain resting sharks (Taylor, 1992). In addition, evidence challenging Porges' proposition that myelinic fibers of the vagus nerve are solely present in the ambiguous nucleus of mammals has surfaced, with reports indicating the presence of myelin in the fibers of cardiac nerves in fish (Monteiro et al., 2018). Furthermore, the amplitude of RSA may be influenced not only by vagal brake activity but also by concurrent changes in sympathetic activity. There are situations in which the coupling between RSA and vagal control over cardiorespiratory rhythm may be disconnected. In light of this evidence, Grossman has suggested that RSA might be an indirect reflection of cardiac vagal tone but is not synonymous with it, thus at least apparently questioning a fundamental tenet of the polyvagal theory.

More recently, a meta-analysis examined the correlation between vagal tone, as indicated by RSA, and the risk of psychopathology. The study analyzed the effect sizes from 37 research endeavors involving more than 2000 participants (Beauchaine et al., 2019). While substantial evidence links low baseline RSA levels to clinical conditions such as internalizing disorders (e.g., depression), externalizing disorders (e.g., conduct disorder), schizophrenia, and sociopathy (Koenig et al., 2016; Shader et al., 2018), the 2019 meta-analysis revealed only a partial effect in the association between vagal tone reactivity and psychopathology. The observed high heterogeneity across studies suggests potential moderators influencing this association. Notably, the link between vagal regulation and psychopathology appears more pronounced in female subjects (Yaroslavsky et al., 2013). Furthermore, variations in key methodological aspects, such as the duration and mode of collecting RSA baseline values, stimulus conditions, and tasks measuring vagal reactivity, contribute to the observed differences among studies. Ultimately, the association between RSA and psychopathology might be more discernible concerning the capacity to restore homeostatic balance and, consequently, recover basal levels of vagal tone (Rottenberg et al., 2007).

In unison

While Porges' theory has faced scrutiny for its evolutionary assumptions and exclusive application to the mammalian model, it has found numerous applications over the years, particularly in the field of developmental psychobiology.

The study conducted by Bazhenova and colleagues (2001) marked a pivotal moment by demonstrating that alterations in a child's emotional state regulation can manifest in vagal tone oscillations. Through a modified version of the still-face paradigm, they revealed that negative emotional states corresponded to decreases in RSA, while positive emotional states correlated with reduced vagal tone and increased RSA values. Subsequent studies, such as those by Ritz et al. (2012), further expanded on Bazhenova and colleagues' findings. Moore and Calkins (2004) discovered that even three-month-old infants exhibited decreased vagal tone during the maternal face immobility phase. Children lacking this physiological response displayed lower positive emotionality during parental interactions and diminished behavioral synchrony with their mothers. However, it is important to note that young children may exhibit variations in vagal system reactivity, with some suppressing vagal tone as expected, while others show non-responsiveness or an increase in RSA (Moore & Calkins, 2004). A 2014 study by Montirosso et al. found that four-month-old infants exposed to the still-face paradigm twice, two weeks apart, exhibited habituation in the second session only if they had significant RSA suppression in response to the maternal immobile face during the initial exposure.

In infants at six months of age, RSA was assessed during the still-face paradigm with their mothers (Moore, 2010). Offspring of parents reporting elevated levels of intra-parental conflict exhibited reduced baseline RSA levels throughout the procedure. Moreover, they displayed diminished responsiveness and instances of vagal brake removal during interactions with their mothers. These findings align closely with Porges' neuroception concept, suggesting that exposure to adverse events, such as parental conflict in this case, may lead to dysregulation in children's ability to perceive social situations as safe and protective (Porges, 2015). In addition, Provenzi and colleagues (2015b) demonstrated that the interaction quality with parents plays a role in shaping the response to the still-face paradigm in four-month-old infants. Infants from dyads characterized by high dyadic repair during the play phase exhibited a significant reduction in vagal tone during the still-face phase and a greater RSA recovery in the reunion phase compared to infants in dyads with lower rates of dyadic repair. This underscores the notion that the child's vagal tone regulation is influenced, in part, by the social context's quality, indirectly supporting the idea that autonomic regulation contributes to the development of affiliative behaviors in contexts perceived as protective and supportive.

Additional confirmation of the impact of early parent-child relationship quality on the development of the vagal regulation system comes from the investigation of behavioral and physiological synchrony processes in mother-child and father-child dyads conducted by Ruth Feldman (Feldman & Eidelman, 2007). This study observed and assessed dyads comprising mothers or fathers and full-term infants for interaction quality and behavioral synchrony when the infant was three months old. The study found a significant association between behavioral synchrony in paternal and maternal interactive dyads and the regulation of the child's vagal tone during infancy. The authors propose that a more mature vagal system at birth may offer infants enhanced neurobiological readiness to regulate the intricate interactive dynamics of face-to-face exchanges with parents in the initial months of life.

In the case of older children, specifically those aged four and five living in poverty, Wolff and colleagues (2012) gauged vagal reactivity in response to various cognitive, physical, and emotional stressors. Subsequently, these subjects encountered a second set of stressful events while being randomly assigned to conditions of either having or lacking social support, and a sympathetic response index was recorded. Children who exhibited effective suppression of vagal tone were able to derive benefits from social support, displaying a more adaptive sympathetic response compared to children possessing similar vagal regulation capacities but lacking social support. Moreover, factors such as exposure to frequent adverse events during the parent's childhood and stress experienced during pregnancy have been identified as predisposing elements for vagal dysregulation in four-month-old infants when subjected to the still-face paradigm (Gray et al., 2017). Those infants displaying greater suppression of vagal tone at one month demonstrated increased dysregulation at three years, but this effect was prominent only in the presence of elevated parental stress. This again indicates that the enduring impact of early parasympathetic system regulation is, to some extent, influenced by the social context and the quality of parental behavior (Conradt et al., 2016).

5 The culprit of stress

Life exists owing to the possibility of maintaining a complex dynamic balance between the demands that come from inside and outside and the regulation of the oscillatory processes of the organism. Under favorable conditions, individuals can invest in pleasurable activities that increase their emotional well-being and promote their psychological, behavioral, and cognitive growth. However, the messy development processes of a non-linear dynamic system involve—by definition—exposure to environmental conditions that may place demands on the individual that exceed their ability to cope with them. We generally call these experiences *stress*. Our body has developed over the course of evolution specific strategies for responding to and regulating states of stress. An evolutionarily archaic physiological mechanism has already been introduced in Chapter 4 and is represented by the vagal brake. A second strategy concerns the activity of a neuroendocrine axis distributed in our body, involving both the central nervous system and the periphery of our body. The end product of this system is a hormone—cortisol—which is generally identified as the hormone responsible for stress (Field & Diego, 2008). However, before understanding how this neuroendocrine axis works and what implications it has for the adaptive development of the child in their physical and relational environment, it is important that we agree on what we mean by the term "stress" from a psychobiological point of view.

Let's stress it out

The earliest recorded mention of a concept akin to stress dates back to Empedocles, approximately 450 years before Christ. Empedocles perceived all matter as a collection of fundamental elements existing in harmonious equilibrium. His contemporary, Hippocrates, expanded this notion to humans, viewing this harmonious balance as a prerequisite for health. Conversely, varying degrees of disharmony were considered predisposing factors for the development of pathological states or diseases. Hippocrates was also the first to discuss adaptive bodily responses, what we now recognize as physiological stress. The Epicurean idea of balance aligns with the definition of homeostasis, a term coined by Claude Bernard in the nineteenth century to denote a state in which the organism is stable and shielded from forces that could lead to its destruction or disintegration. It's noteworthy that

DOI: 10.4324/9781003479314-7

Bernard argued that a living system is stable, to some extent, precisely because it is adaptable. In other words, a slight instability is a necessary condition for genuine stability. This concept foreshadows what Sander expresses in a more organized and organic manner (refer to the first chapter of this volume).

In the 1930s, Hans Selye emerged as one of the earliest scholars to employ the term "stress" in the manner understood today in psychobiology. Drawing inspiration from physics, where it denoted force on a surface, Selye adapted the term to signify the impact of forces occurring at various levels within an organism. In alignment with the Hippocratic tradition, Selye recognized that individuals facing serious pathologies often manifested common bodily symptoms and signals such as gastrointestinal issues, weight fluctuations, alterations in appetite, and depression. This Hungarian-born Canadian researcher substantiated these effects in an animal rat model, revealing that diverse types of stressful events not only altered behavior but also impacted the neurobiology of the animals. Based on these findings, he posited that living organisms possess non-specific mechanisms to adapt to various sources of adverse influences. Selye introduced the term "general adaptation syndrome" to describe a standardized physiological response occurring in three stages. Initially, there would be an alert reaction mediated by the release of epinephrine and glucocorticoids, capable of disturbing and subsequently restoring a state of calm in the organism. Following this, the individual would enter a phase of resistance, striving to achieve an optimal state of adaptation. However, if stress persisted over an extended period, the individual would undergo a phase of exhaustion, resulting in the cessation of the adaptive response and paving the way for long-term consequences on the individual's health and well-being (Selye, 1936). It is noteworthy that many components of the neuroendocrine system's stress response were only discovered and described relatively recently. Although Harris hypothesized in 1948 that the hypothalamus was partially controlled by the pituitary, the role of cortisol in stress has been extensively studied only in the past 60 years, and corticotropin-releasing hormone (CRH) was isolated by Vale and colleagues in 1981 (Chrousos et al., 1988).

We now understand that not all types of stress elicit the same stereotypical response described by Selye (McEwen, 2000). In a broader sense, stress can be defined as a real or subjectively perceived threat to the physiological and/or psychological integrity of an individual, leading to specific behavioral, physiological, and neurobiological responses (McEwen, 2000). This definition underscores the three elements inherent in the experience of stress. The first element pertains to the threat or, more generally, the input—the signal perceived as dangerous. This element is commonly referred to as the "stressor." The second element involves the processing of the stressor by various systems, encompassing biological, physiological, neural, or behavioral mechanisms. These systems intricately involve the regulation of the autonomic nervous system and the activation of the organism's neuroendocrine response. These systems are widely distributed throughout our body, emphasizing the complex and systemic nature of stress processing (Levine, 2005). The third element is the observed behavioral response following stressful events. It is crucial to emphasize that the experience of stress comprises components of different natures that are not necessarily concurrent in time and intensity (Gunnar et al., 1981; Selye, 1975). In addition, responses to stressful events and life disruptions can be categorized based on their short-term (acute stress) or long-term (chronic stress) nature. Furthermore, the process of stress regulation can be delineated into two phases: a reactivity phase and a recovery phase. Reactivity pertains to how individuals prepare for and respond to the stressful event, while recovery denotes the ability to restore a state of balance or homeostasis once the disturbance has subsided (Haley & Stansbury, 2003).

A neuroendocrine cascade

The organism's stress response primarily involves the activity of two components—neurophysiological and neuroendocrine: the sympathetic/parasympathetic system (refer to Chapter 4) and the hypothalamic-pituitary-adrenal axis (HPA axis). The autonomic nervous system governs the activity of numerous organs and endocrine glands, comprising two main branches: the sympathetic system and the parasympathetic system. While the sympathetic system aids in coordinating and initiating the response to stimulation, the parasympathetic system plays the opposite role, primarily focused on reducing overall organismal activation. The sympathetic-adrenergic system, a subcomponent of the sympathetic nervous system, contributes to increasing heart rate and promoting the release of catecholamines such as epinephrine and norepinephrine at the adrenal medulla level. The rapid release of catecholamines facilitates the body's immediate response to stress, triggering the fight or flight reaction.

In contrast, the HPA axis operates at a slower pace, involving a cascade of neuroendocrine reactions (Joëls et al., 2008). Following a stressful event, the hypothalamus releases CRH, which, in turn, prompts the release of adrenocorticotropic hormone (ACTH) in the anterior pituitary region. ACTH, in humans, leads to the production of several steroid hormones, primarily cortisol, by the outer cortex of the adrenal gland. Moreover, the HPA axis can regulate the levels of cortisol in the bloodstream, decreasing them once the environmental disturbance concludes. As cortisol levels rise in response to an acute stressor, feedback mechanisms kick in, reducing cortisol secretion by suppressing the synthesis of CRH and ACTH. These feedback mechanisms operate relatively slowly and can exert their effects over a period ranging from 5 to 40 min (Keenan et al., 2001).

The acute activation of the HPA axis, marked by the release of cortisol in response to individual, specific stressful events, triggers various effects within the body. These include the mobilization of energy to the muscles, heightened cardiovascular tone, stimulation of the immune system, inhibition of reproductive physiology and appetite, facilitation of certain cognitive processes, and the utilization of glucose in specific areas of the brain (Sapolsky et al., 2000). Moreover, cortisol is not exclusively produced in response to stress but is released in phases throughout the day to maintain essential basal hormone levels necessary for the body's metabolic functioning. Basal secretion follows a circadian rhythm: it remains low during inactive sleep phases, increases around waking, and peaks in the morning hours. This basal secretion occurs pulsatile, with the amplitude of these pulses typically starting low during the early circadian cycle (during periods of inactivity), increasing toward the onset of activity, and gradually decreasing thereafter, forming the characteristic sinusoidal profile of the diurnal regulation of the HPA axis.

Two categories of steroid hormones are synthesized: glucocorticoids and mineralocorticoids. In humans, cortisol serves as the primary glucocorticoid, while aldosterone functions as the main mineralocorticoid (de Kloet et al., 1998). Although cortisol plays a more direct role in the stress response, both glucocorticoid receptors (GR) and mineralocorticoid receptors (MR) contribute to the adaptive stress-regulating response. GRs exhibit higher affinity for cortisol and are predominantly activated during stressful situations or circadian peaks (Joëls et al., 2008). While direct exploration of the expression of genes encoding GR and MR in humans is not feasible, studies in animal models have revealed that GRs are distributed across various brain regions (Patel et al., 2000), encompassing not only the hippocampus but also the prefrontal cortex. These regions are implicated in the formation of working memories and long-term memories, underscoring the broad involvement of the HPA axis in the establishment of new learning. In mice, corticosterone, an analogue of cortisol, has been found to either facilitate or impede the formation of new memories depending on the type of memory process considered (encoding, consolidation, retrieval) and the cortisol levels produced during the experience (de Kloet et al., 1999).

The regulation of the HPA axis involves retroactive mechanisms in which the endocrine outputs, represented by cortisol, exert feedback through both a direct and an indirect pathway. In the direct route, cortisol molecules bind to GRs located in components of the HPA axis, such as the paraventricular nucleus of the hypothalamus or the anterior pituitary, leading to the cessation of secretion activity. Sensitivity to this feedback mechanism varies among individuals and is largely dependent on the availability or density of GRs. The indirect pathway operates through the action on GRs and MRs in various structures, with a specific role attributed to the hippocampus. The hippocampus exerts an inhibitory effect on the HPA axis, particularly through the excitatory projection of hippocampal neurons onto inhibitory GABAergic neurons near the paraventricular nucleus of the hypothalamus. Elevated hippocampal activity progressively suppresses the HPA axis. The predominant activation of hippocampal MRs enhances hippocampal excitability, leading to HPA axis inhibition. On the other hand, GR activation suppresses hippocampal function, disinhibiting paraventricular neurons and elevating neuroendocrine activity. Given the distinct nature of activation and inhibition, the balance of MRs and GRs is pivotal in comprehending the role played by the HPA axis in stress response, as well as in learning and memory processes (de Kloet et al., 1998).

Everyday resilience

Many of the investigations that have paved the way for understanding the HPA system and its involvement in stress regulation and memory have utilized the administration of painful stimuli, such as physically invasive sampling or procedures

(Gunnar et al., 1991; Lewis & Ramsay, 1994). These stressors typically induce a substantial increase in cortisol secretion, which diminishes only after a period of habituation. Recent studies have increasingly employed stresses with greater ecological validity that may be more suitable for children's age. These stressors are of a mild or moderate social nature and are not consistently linked to significant increases in cortisol secretion, such as bathing (Albes et al., 2008), behavioral reflex assessments (Keenan et al., 2002), motor learning procedures (Haley et al., 2006), or disruptions in mother-child communication (Provenzi et al., 2016). During these procedures, there has been a noticeable individual variation not only in the magnitude but also in the direction of the HPA axis response, suggesting that while some children exhibit an increase in salivary cortisol levels, others may display an opposing pattern (Gunnar et al., 2009).

From the perspective of developmental psychobiology, the examination of HPA axis regulation becomes particularly intriguing when considering social stressors. While the investigation of neuroendocrine reactivity to physically oriented stress stimuli, such as thermal shock or pain, has significantly enhanced our comprehension of HPA axis functionality over the years, real-life conditions for many individuals rarely involve such circumstances. Except for initial visits to the pediatrician, most individuals experience life characterized by various stressors, stemming mostly from the physiological consequences of the regulatory processes inherent in nonlinear dynamical systems—the inevitable interactive disruptions described by Tronick. Understanding how the body responds to these events at the neuroendocrine level holds special significance. It can be hypothesized that it is precisely through the experience of regulating these interactive disruptions, from the early stages of development and within the interactive narrative with caregivers, that we learn to develop adaptive strategies for handling future stressful events in life. This aligns with the concept of "resilience of everyday life" proposed by DiCorcia and Tronick (2011).

The term "resilience" is frequently employed to describe an individual trait exhibited by those who navigate challenging circumstances without succumbing to the most adverse outcomes, even when exposed to early adversities (Cicchetti et al., 1993). Tronick, in fact, proposes viewing resilience as an epiphenomenon—an emergent characteristic of non-linear dynamical systems—resulting from dyadic parent-child regulatory processes. It is estimated that between three and six months of age, children experience states of stress approximately 10% of the time, exhibit positive affectivity for around 13% of the time, and even during interactions with adults, they may undergo physiological transitions into states of frustration or stress for about 3% of the time (Weinberg et al., 1999). These are typically transient states, often lasting less than 5 min, and children can self-regulate through behaviors like finger sucking, engaging with an object, or averting their gaze. However, a child's ability to cope with stress is not entirely within their control. The active involvement of parents in managing the child's stress states through appropriately attuned and contingent interventions is crucial for emotional regulation. Nevertheless, this contingency is far from being the norm in the interactive dynamics of the parent-child dyad, and interactive disruptions are largely unavoidable.

In this scenario, one could postulate that the ability to navigate daily stressors contributes to the development of the child's regulatory skills and lays the groundwork for the long-term programming of behavioral and neurobiological systems related to stress reactivity. Building upon Hofer's exploration of hidden regulators (Hofer, 2006) and Tiffany Field's insights into the co-regulation of mother-child biological rhythms (Field, 1994), Tronick suggests viewing the daily encounter with psychosocial stress as an ongoing learning experience that can shape emotional regulation development, extending its impact beyond acute or traumatic adverse events. He draws an analogy with marathon training: runners typically don't cover the full marathon distance each day in preparation; instead, they run specific distances daily and gradually increase the distance. Similarly, a child engages in small, transient co-regulations of daily stressors with the caregiver, progressively developing adaptive coping skills until they are well-prepared for larger challenges.

I've previously introduced Tronick's mutual regulation model in the early chapters of this volume. In this model, dyadic interaction inherently involves moments of mismatch or interactive breaks, which the dyad resolves by negotiating the organization of their interactive contributions in a moment-to-moment fashion. Central to Tronick is the repair process, through which the dyad restores a state of alignment. Repeated experiences of interactive repair—connections between organism and environment—contribute to the child's development of adaptive regulation skills for future stressful situations, utilizing integrated behavioral and psychobiological strategies—connection within the organism. An illustration of this concept is found in the examination of interaction and co-regulation processes within the still-face paradigm. A 2016 meta-analysis by Provenzi and colleagues revealed that young children exhibit a significant HPA axis response to disruptions in maternal communication—subtle, age-appropriate disturbances in interaction. Specifically, repeated exposures to these disruptions appear to result in a measurable increase in cortisol secretion, although substantial individual differences exist (Provenzi et al., 2016c). Therefore, it is plausible to suggest that a child develops resilience skills through recurrent exposure to minor, fleeting stressful events that they learn to regulate independently and with the interactive support of the parent. This gradual exposure to stress in small doses could allow the HPA axis to accumulate effective repair learnings, and the sensorimotor cortex could encode memories of interactive, intra-organismic integrations, serving as a model for subsequent adjustments to stressful events throughout childhood and into adulthood.

Pump up the volume or shut the system down?

While the activation of the HPA axis plays a role in adaptively managing stressful or socio-emotionally challenging situations, persistent or prolonged engagement of this neuroendocrine system can lead to neurotoxic effects with enduring implications for health (Heim et al., 2002; Alink et al., 2012). Given that the brain is a primary target of HPA axis activity, it comes as no surprise that the impact on mental health is substantial. Even in infants, whose neuroendocrine system is partially mature and still developing, exposure to high levels of early stress doesn't exempt them from the risk of developing maladaptive outcomes (Gunnar & Herrera, 2013). However, the literature provides conflicting findings, encompassing both hypo- and hyper-regulation patterns of the HPA axis (Strüber et al., 2014). It is crucial to note that these consequences extend beyond baseline cortisol levels and encompass the potential for an altered response—be it inhibited or excessive—of the HPA axis to subsequent stressful events (Heim et al., 2000).

Early life experiences have the potential to shape the functioning of the HPA axis, influencing its ability to regulate stress both during childhood and into adulthood. In animal models, this programming serves the adaptation of individuals to their specific environments and the challenges they encounter (Del Giudice et al., 2011; Ellis & Boyce, 2008). As early as the 1950s, Levine (1957) demonstrated that manipulative experiences in the early stages of life could have enduring effects on the adrenocortical system in mice. During handling, newborns are briefly separated from their mothers once or twice a day and placed in small containers or held by the experimenter. Those exposed to manipulation during the first weeks of life exhibited a diminished corticosterone secretion profile in response to subsequent stressful events in adulthood. However, the duration of maternal separation appears to be a critical factor. The typical manipulation procedure involves a separation lasting 3–15 min (Meaney et al., 1994). Prolonged separation exceeding three hours per day during the initial two weeks of life results in a distinct response, with animals developing hypersensitivity to stress and elevated cortisol levels (Plotsky & Meaney, 1993). Other studies have presented conflicting findings, reporting both increases and decreases in baseline levels and reactivity of HPA axis activity (Feng et al., 2011; Ladd et al., 2000). The age of the child at the time of separation and manipulation may play a pivotal role, with very early separations, within the first days of life, associated with hypersecretion of corticosterone, while later separations, within the initial 20 days of life, potentially fostering a profile of neuroendocrine hyporesponsiveness (Enthoven et al., 2008; van Oers et al., 1998).

In human studies, diverse findings have been reported regarding diurnal cortisol production, with some indicating an increase (Halligan et al., 2004) and others revealing a pattern of decreased basal cortisol production (Lupien et al., 2001). In a study involving two-year-old orphans, Carlson and Earls (1997) found a significantly reduced level of salivary cortisol in the morning compared to children raised in families. Moreover, these orphaned children exhibited an atypical absence of the expected decline in cortisol levels throughout the day, displaying a flattened circadian pattern (Bevans et al., 2008). Investigations into the relationship between early adverse experiences and baseline regulation patterns of the HPA axis in adults have presented conflicting findings, documenting both increases (Gonzalez et al., 2009) and decreases (Meinlschmidt and Heim, 2005) in basal cortisol secretions upon awakening. These discrepancies may be attributed, in part, to variations in observational procedures (McCrory et al., 2010). The nature and intensity of stressors can vary concerning age. A study by Van der Vegt et al. (2009) explored the enduring connection between early maltreatment experiences and baseline cortisol concentrations in over six hundred adults who had been adopted as children. Individuals subjected to severe neglect or abuse during childhood exhibited lower morning cortisol levels, while those experiencing neglect showed a flattened circadian curve. Moderate levels of abuse were associated with elevated cortisol levels and a more pronounced circadian regulatory curve. This study underscores how distinct types and severity levels of early stress can elicit divergent effects on HPA axis programming, with potential mediation by the child's behavioral characteristics. Cicchetti and colleagues (2010) demonstrated that children subjected to early physical or sexual abuse displayed a flattened circadian cortisol regulation curve compared to non-maltreated children, but this effect was evident only when accompanied by high levels of internalizing behavior problems. Early adversity may also lead to altered reactivity profiles of the HPA axis. Prenatal exposure to maternal stress or depression has been associated with hypo-responsiveness profiles of the HPA axis in the child post birth (Weinstock, 2008). Conversely, Heim and colleagues (2000) reported increased sensitivity of the HPA axis to stressful stimuli in children with a history of maltreatment, while other researchers found an association between early adverse experiences and a reduced cortisol response later in the child's life (Carpenter et al., 2007; MacMillan et al., 2009; Shenk et al., 2010).

What makes a good rat mama

Undoubtedly, maternal behavior stands out as one of the environmental factors with significant potential to shape the development and functioning of the HPA axis. In rodent studies, mothers of young subjected to manipulation by experimenters spend an equivalent amount of time engaged in caring activities for their litter compared to mothers of non-manipulated young. However, manipulated mothers exhibit an increased frequency of behaviors such as licking-grooming (LG) of the young, actively promoting breastfeeding, and assuming an arched position for heating the offspring (arched-back nursing, ABN) (Liu et al., 1997). Consequently, Meaney's research group embarked on investigating how natural physiological variations among a group of mothers could impact the HPA axis development in rodents. Their findings revealed that offspring raised by mothers displaying a high frequency of LG-ABN behaviors exhibited reduced levels of ACTH and corticosterone in response to laboratory stressors during adulthood, in contrast to their counterparts cared for by mothers with low LG-ABN frequencies. No differences were observed in baseline levels. Liu and colleagues (1997) concluded that variations in individual maternal behavior might be linked to divergent neuroendocrine stress responses across generations.

The enduring programming of the HPA system, specifically the production of hippocampal receptors, appears to be, at least partially, influenced by the environment, particularly parental behavior. As detailed in chapter seven of this volume, the mechanisms facilitating this process seem to be of an epigenetic nature—functional alterations of DNA that modify the extent of gene expression. For the purposes of this chapter, it suffices to understand that one specific epigenetic process, DNA methylation, seems highly responsive to variations in maternal care in mouse offspring. Notably, the progeny of low-LG-ABN mothers exhibits elevated levels of methylation in the gene responsible for hippocampal GR receptors, an effect not observed in the offspring of mothers characterized by a high frequency of LG-ABN behaviors (Weaver et al., 2004). This mechanism appears to be further mediated by serotonin's action, synthesized in response to maternal LG behaviors, facilitating the gene expression of GR receptors in the baby's hippocampus (Meaney, 2010).

Given the hippocampus's inhibitory role in the HPA axis (de Kloet et al., 1998), Meaney's findings might initially appear surprising. Theoretically, an increase in the expression of hippocampal GR, which inhibits hippocampal activity, should lead to heightened HPA axis activity. However, this contradicts the results of Liu and colleagues (1997), who observed an upregulation of GR expression in the dorsal hippocampus. Studies on individuals with brain lesions indicate that the dorsal hippocampus inhibits basal HPA axis activity, while the ventral region modulates stress reactivity and glucocorticoid secretion (Jankord & Herman, 2008). Therefore, an elevation in GR expression in the dorsal hippocampus would theoretically reduce activity in this area and increase baseline levels. Although Liu and colleagues (1997) found no baseline differences in young mice from low and high LG-ANB mothers, measurements were taken shortly after stress exposure. Given the circadian nature of basal glucocorticoid secretions, assessing cortisol levels at a specific time, such as upon waking, would have been preferable. Consequently, it cannot be ruled out that there were indeed baseline differences in HPA axis activity between the two groups. As a result, an increase in dorsal hippocampal GR expression due to high maternal LG-ABN may lead to elevated basal corticosterone secretion and a concurrent reduction in stress reactivity.

Another contentious aspect revolves around the role of serotonin in the link between the quality of care and HPA axis programming. According to Meaney's model (2004), elevated levels of LG would foster serotonin production in the hippocampus and greater expression of GR receptors (Laplante et al., 2002). However, there is evidence suggesting an association between high and chronic stress levels and increased serotonergic activity in the hippocampal region (Storey et al., 2006). Could it be plausible that an increase, within species-specific variations, in licking and grooming behaviors indicates greater stress for the mother or offspring? In rodents, mothers exhibiting high levels of anxious behaviors display an extremely protective caregiving style, including a heightened frequency of LG-ABN behaviors (Bosch & Neumann, 2012). Considering this perspective, one might question whether the surge in LG-ABN observed in previously manipulated rodent mothers is a compensatory expression on the part of the parent. Macrì and Würbel (2006) proposed a similar interpretation, suggesting that variation in the quality of maternal care may be an incidental epiphenomenon, rather than a causal factor, involved in the development of the HPA axis in the offspring.

Too much and too little

We are aware that parental nurturing behavior can, to some extent, mitigate the rise of cortisol in early life by engaging the oxytocinergic system. Oxytocin is released in response to various physiological and environmental stimuli, such as birth, sucking, sexual activity, and different forms of stress. It plays an inhibitory role on basal cortisol levels and corticosterone reactivity. In humans, intranasal administration of oxytocin has been shown to reduce cortisol levels in response to stressful socio-emotional stimuli (Ditzen et al., 2009; Grumi et al., 2021).

Maternal behavior is linked to oxytocin release in the maternal brain, and elevated levels of LG-ABN facilitate the gene expression of oxytocin receptors in various brain regions (Champagne et al., 2001). Non-human primates raised separately from their mothers have been found to have lower concentrations of oxytocin than those cared for by their mothers at the same age (Winslow et al., 2003), a pattern also observed in women with a history of abuse (Heim et al., 2009). Oxytocin concentrations in response to physical contact with the mother increase in children raised in families, whereas this effect is not observed in children raised in orphanages (Fries et al., 2005). It is noteworthy that, in addition to acting as an antagonist of glucocorticoid secretion, oxytocin facilitates the central release of serotonin.

Building on these findings, Strüber and colleagues (2014) put forth an intriguing hypothesis suggesting two distinct programming paths for HPA axis functionality. This speculative proposal seeks to reconcile and address the controversies previously highlighted regarding the effects of early adverse events on HPA axis hyporesponsiveness and hyperreactivity. The authors posit that in cases where early adversity lacks sufficient maternal care – either due to separation from the mother or exposure to maltreatment or neglect – the experience of stress becomes linked to chronic and repetitive HPA axis activity, frequent cortisol secretion, and potential neurotoxic effects. This may induce a habituation phenomenon, programming low levels of hippocampal expression of GR receptors (Arabadzisz et al., 2010). Given the inhibitory role of the hippocampus on the HPA axis, an increase in the expression of hippocampal GR receptors would heighten glucocorticoid secretion. Consequently, repeated early adverse experiences without adequate maternal care could foster the emergence of HPA axis hyporesponsiveness. Conversely, when the mother is present, the experience of stress would be associated with the activation of the oxytocinergic system. This activation would inhibit cortisol secretion and enhance serotonin release, resulting in an increase in the density of hippocampal GR receptors. In turn, this would lead to decreased activation of the hippocampus, programming future hyperreactivity to stress in the HPA axis.

Paying the consequences

We have explored how the HPA axis not only enables responses to acute stressful stimuli but also possesses the capacity for learning from early adverse experiences. Such learning can either foster the development of resilience skills (DiCorcia and Tronick, 2011) or contribute to dysregulation—manifesting as hyporesponsiveness or hyperreactivity—within the neuroendocrine system. However, the question arises: How does this phenomenon occur?

While Selye's model implied that negative health consequences stemmed from the failure of adaptation leading to a phase of exhaustion, McEwen (2000, 2006) proposed an alternative perspective. According to McEwen, the same agents involved in the stress response can induce neurotoxic effects, giving rise to long-term iatrogenic consequences. McEwen suggests that, to maximize survival chances, our biological processes have evolved to maintain stability and adaptation, a concept known as allostasis. Allostasis relies on the capacity to detect changes in the external and internal environment and activate specific adaptive responses. Three closely integrated systems—the nervous, endocrine, and immune systems—offer the potential for developing allostatic states. Because these systems are intricately connected, stimulation of one allostatic system affects them all. Psychosocial stress, for instance, is processed by a neurobiological network involving the thalamus, sensory cortex, and amygdala. The amygdala's activity is regulated by two inhibitory afferents: the hippocampus, which influences through learning and memory processes based on prior experiences, and the prefrontal cortex, which engages in executive functions, attention, and meta-cognition processes. The amygdala, in turn, activates the locus coeruleus, triggering heightened alertness and attention to the environment, resulting in the activation of the sympathetic nervous system— a response akin to Selye's fight-or-flight response. Simultaneously, inflammatory processes and an immune system response are initiated to prevent extensive tissue damage. Additionally, the amygdala stimulates the hypothalamus to initiate a neuroendocrine response involving the HPA system, leading to the circulation of catabolic activity aimed at mobilizing energy at the peripheral level. This system is indeed complex and non-linear, with various systems collaborating to maintain coherence and adaptation across space and time (McEwen, 2006).

As mentioned earlier in relation to the HPA axis, chronic or repeated activations by psychosocial stressors can result in the prolonged activation of these allostatic systems. Consequently, allostatic load occurs, or in more severe cases, allostatic overload. McEwen uses this term to describe the cost the body pays to find a balance between adaptation and coherence in the face of particularly stressful or repeatedly adverse environmental situations. McEwen applies this concept specifically to a few crucial physiological subsystems—including pH, body temperature, glucose levels, and oxygen saturation—that are vital for life and must be maintained within specific ranges or very narrow threshold values. On the other hand, the concept of allostasis pertains to systems in which stability is achieved through changes and adaptations. The primary mediators of allostasis include, but are not limited to, hormones participating in the neuroendocrine cascade of the HPA axis, catecholamines, and cytokines. Allostasis helps clarify the ambiguity of the term homeostasis by distinguishing between systems essential for maintaining life and those essential for adaptation. When control threshold values deviate beyond the limits, allostatic adaptations occur. An allostatic state results in an imbalance of primary mediators, reflecting excessive production of some and insufficient production of others. Examples include chronic hypertension, a flattened profile of circadian cortisol regulation in depression, and chronic elevation of inflammatory cytokines in chronic fatigue syndrome. Allostatic states can only be sustained for a limited period. If environmental demands or stress increase or persist over time, symptoms of allostatic load become apparent. In animals raised in captivity in zoos, these symptoms manifest as abdominal obesity. Therefore, allostatic states refer to altered activity levels of primary mediators that are maintained for an extended duration, potentially causing wear and tear on neurological and neu-roendocrine regulatory systems. The terms allostatic load or allostatic overload denote the cumulative consequences of allostatic states, serving as a side effect of our adaptive capacity.

From this standpoint, the alarm phase or Selye alert can be reconsidered as an attempt at adaptation—allostasis—where glucocorticoids and epinephrine (along-side other mediators) are released into circulation to facilitate adjustment to the stressor. The resistance phase corresponds to the protective effects of this adapta-tion and the efficiency of the body's implemented mechanisms. However, the pro-longed and/or frequent repetition of the alarm phase can lead to an allostatic load. This contrasts with the Selye depletion phase, representing the inevitable wear and tear resulting from repeated exposure to allostasis mediators. Essentially, it serves as a biological or neuroendocrine manifestation of "too much," akin to envisioning the HPA axis as a short blanket attempting to strike a balance between the need for adaptation and the preservation of system functionality and coherence.

It is crucial to highlight that in children, many of these systems are not yet fully mature. The human brain undergoes continuous maturation until adulthood: the vol-ume of white matter increases from childhood to adolescence, while the gray matter exhibits an inverted U-shaped trajectory during the first two decades of life, influenced by synaptic pruning processes and the evolutionary events of myelination. These pro-cesses are not uniform and affect different brain areas at various times and rates (Casey, Galvan, Hare, 2005). The sensorimotor cortex matures in the initial three years, aiding in the acquisition of sensory and motor integrations. The parietal and temporal cortex matures within the first ten years, facilitating the development of language and spatial skills. Finally, the prefrontal cortex reaches maturity well beyond puberty, supporting the mature development of executive functions. The HPA axis undergoes a maturation of the circadian cycle, featuring two daily peaks up to the initial two months, the emergence of a diurnal peak in the initial four months, and the development of a mature regulation similar to that observed in adulthood around four years of age (Gunnar & Donzella, 2002). Baseline cortisol levels also gradually decrease over the initial year of life (Tollenaar et al., 2010), exhibiting substantial individual differences. The immune system also undergoes age-dependent maturation: newborns are shielded from infection by maternal anti-bodies transferred via the placenta and milk. However, despite the presence of an innate immune response, it can only recognize a limited number of pathogens, and during early development, colonization of the skin, gastrointestinal tract, and lungs further facilitates the development of the acquired immune system.

The repercussions of allostatic load in children are diverse, impacting all the various systems involved in allostasis. As discussed earlier, these effects can mani-fest as dysregulation in the HPA axis. In the realm of the central nervous system, the prefrontal cortex may experience dendritic wear and tear, leading to a reduction in attentional skills and the ability to extinguish a fear response. The amygdala's dendritic growth may be inhibited, diminishing the efficiency of behavioral mani-festations of emotional states, while a reduction in hippocampal volume can result in learning and memory deficits. On the immune front, prolonged inflammatory states may induce a gradual inefficiency of the HPA axis (Glaser et al., 1999). Further insights into the repercussions of early adversity will be presented in the third section of this volume, highlighting how allostatic load may be at play in the cases of children exposed to prenatal stress (Chapter 8), preterm births (Chapter 9), and instances of maltreatment or abuse (Chapter 10).

Parenting as a protective buffer

In mice, the final days of gestation and the initial hours after birth are marked by a decrease in glucocorticoid production, maintaining a relatively high baseline. Corticosterone concentrations decrease during the first two postnatal days and remain low until approximately 14 days. Stressors that would typically activate an HPA axis response become less effective during this refractory period. According to Megan Gunnar, this period of reduced HPA axis responsiveness is also evident in humans (Gunnar & Cheatham, 2003; Lupien et al., 2009). In 1996, Gunnar and colleagues demonstrated that children aged six to fifteen months may experience a decline in response to physical stress, such as vaccination puncture, previously observed between two and four months. A decrease in basal cortisol values is also noted during the same period (Lupien et al., 2009).

Experiments conducted with rodents have revealed that specific maternal factors can influence the refractory period. Notably, the quality of interaction between the mother and the baby can modulate neuroendocrine activation in response to postnatal stress, leading to the inhibition of cortisol secretion and resulting in a state of hypo-responsiveness. A separation from the mother for eight hours on the third day led to a diminished corticosterone response to novel stimuli (Enthoven et al., 2008). Moreover, the low level of corticosterone reactivity persisted over the following 2 days, and baseline ACTH levels remained suppressed even sixteen hours after the reunion with the mother. However, the stress response was maintained and even heightened when maternal separation was combined with an additional stressful stimulus (Daskalakis et al., 2013). Earlier research by Levine and Wiener (1988) had already demonstrated increases in plasma cortisol levels following maternal separation in primates. These increases were less pronounced when other members of the household were present and more pronounced when the child was surrounded by unknown conspecifics.

Under typical circumstances, certain social relationships, particularly early interactions with the caregiver, may potentially restrain or regulate the HPA axis response to stress. In non-human primates, the mother's presence acts as a filtering mechanism, often referred to as a buffer, influencing the neuroendocrine response of the offspring. This buffer allows the young to express stress behaviorally, triggering an appropriate maternal response without concurrent elevations in cortisol. In humans, it is well-documented that early adverse experiences can predispose individuals to develop psychopathological outcomes, partly through alterations in HPA axis function (Heim et al., 1997). Children classified as secure in the strange situation at eighteen months exhibited lower cortisol levels at two or six months (Gunnar et al., 1996). During the strange situation, children with a secure attachment style at one year did not show elevated cortisol levels, while those with ambivalent insecure attachment reacted with significant cortisol secretion (Schieche & Spangler, 2005). Infants with disorganized attachment styles also displayed higher cortisol levels (Hertsgaard et al., 1995). Collectively, these studies suggest that maternal behavior has the capacity to inhibit HPA axis activation in response to disruptive situations. Albers and colleagues (2008) proposed that maternal influence may be particularly evident in the child's ability to restore baseline cortisol levels, rather than in reducing the amplitude of reactivity, aligning with findings from other studies using the still-face paradigm (Provenzi et al., 2015b; Montirosso et al., 2016b).

This parental buffering effect (Gunnar & Donzella, 2002) can be viewed as a facet of the parent-child dyadic system, enabling the child to navigate and express discomfort, effectively communicate emotional states, and seek assistance in stress regulation without overstimulating the HPA system (Gunnar & Donzella, 2002). When we delve into the discussion of parental behavior's role in regulating the HPA axis of preterm infants in chapter nine, we will observe this dyadic co-regulation in action, closely resonating with Hofer's concept of a hidden regulator. However, in contrast to the animal model, where caregiving behavior is more easily categorized through specific behaviors like the LG-ABN pattern, parental behavior and the array of dyadic interactions are considerably more intricate in humans (Provenzi et al., 2018e). Parental sensitivity, for instance, is a nuanced and multifaceted construct encompassing caregiver availability, attentiveness to the child, interpretation of communication cues, timely and appropriate responsiveness in terms of timing and intensity, consideration of the child's perspective, provision of learning opportunities, encouragement of exploration, and more. Different components of parenting behavior may exert distinct effects on the regulation of the child's HPA axis.

Not only stress: The role of HPA in memory and learning

Cortisol assumes a crucial role in stimulating and regulating neural activity within brain regions associated with learning and memory, such as the hippocampus. Research in non-human primates and adults has indicated that consistently elevated cortisol levels may be linked to cognitive impairments (Heffelfinger & Newcomer, 2001). The Trier Social Stress Test (TSST), frequently employed to investigate the HPA axis's impact on human learning processes, requires participants to prepare and deliver a presentation before a panel. In this paradigm, individuals exhibiting a significant cortisol increase often demonstrate less effective performance in cognitive tasks (Kuhlmann et al., 2005). Another paradigm involves immersing the hand in an extremely cold water basin: participants displaying a pronounced cortisol secretion increase tend to exhibit poorer memory performance in verbal tasks compared to those with limited increases (Buchanan et al., 2006).

Thompson and Trevathan (2008) conducted assessments on three-month-old infants involving a maternal separation task and an associative learning task. Infants exhibiting reduced cortisol levels following separation demonstrated enhanced short-term recognition memory. This study was later replicated by the same authors with six-month-old infants (Thompson & Trevathan, 2009). In addition, Van Bakel and Riksen-Walraven (2004) found that insecure children, responding to social stress with elevated cortisol levels, achieved higher scores on an intelligence test. Consequently, it raises the question of whether increases in cortisol levels are linked to cognitive deficits or improved performance, especially in the realm of learning and memory.

Studies in rodents have shown that glucocorticoids—the analog of cortisol in humans—influence long-term potentiation (LTP) formation in hippocampal neurons (see also Chapter 3 of this book). This relationship seems to be due to the activation of two different types of glucocorticoid receptors that were introduced previously: MR receptors, which have a high affinity for glucocorticoids, and GR receptors, which have a lower, about ten times lower affinity (Zhou et al., 2010). As a result, in basal conditions, MR receptors are occupied first, while GR receptors bind cortisol only later, when the concentration increases and the body is facing an environmental condition of stress or is performing a particularly activating task. The relative balance in the occupation of these different receptors would affect memory in adults (de Kloet et al., 1999).

Memory facilitation has been observed in animal models when MR receptors are fully occupied and GR receptors are only partially occupied. Human studies support this hypothesis, potentially resolving previous inconsistencies in the investigation of the link between cortisol and learning (Kirschbaum et al., 1996). In 2006, a study conducted by Ruth Grunau's group in Vancouver aimed to understand the relationship between cortisol and memory in young children (Haley et al., 2006). In this study, three-month-old infants underwent a classical operant learning paradigm. Approximately twenty-four hours later, the infants were tested in a learning task involving a crib, a ribbon tied to their foot, and a hanging carousel above them. Kicking frequency, representing learning, was recorded during different phases. Saliva samples were collected before and 20 min after exposure to the paradigm. Results indicated that 50% of the infants exhibited an increased frequency of calcium index of learning both in the short term and when retested after one day using the same procedure. Furthermore, infants who showed an increase in cortisol levels during the first session demonstrated better memory compared to those with no increase in HPA axis activity or a decrease in salivary cortisol secretion. Specifically, the highest memory performance was observed in infants with an increase in cortisol up to 0.30 ug/dl, while no significant differences were observed in learning for infants with decreased cortisol levels or increases beyond 0.30 ug/dl. These findings support the hypothesis of a quadratic relationship between cortisol increase and cognitive performance in young children, as early as three months, resembling an inverted U-shaped curve described by the Yerkes-Dodson law. These results indirectly confirm the role of the MR/GR ratio in facilitating long-term learning and memory enhancement.

More recently, Montirosso and colleagues (2013) presented additional evidence supporting the role of cortisol in the formation of early memories in children, expanding on Haley's findings to investigate memory for social events. Children as young as four months participated in the Still-Face Paradigm with their mothers twice, separated by 15 days. Saliva samples were collected before the observational procedure in both sessions to establish baseline salivary cortisol values. In addition, three more saliva samples were collected at 10, 20, and 30 min after the conclusion of the still-face paradigm to observe the HPA axis response to this socio-emotional stress. The study aimed to explore whether there were behavioral indicators (negative emotionality display) or neuroendocrine indicators (specifically, salivary cortisol) of any memory in children resulting from repeated exposure to maternal non-responsiveness. To establish a control condition based on age and exposure, a second group of children underwent the still-face paradigm only once at the age of four months.

While there were no behavioral differences observed in the children's responses between the first and second sessions in the experimental group, or when comparing first-time participants from both groups, a noteworthy distinction emerged in terms of neuroendocrine reactivity. Children who exhibited a substantial increase in salivary cortisol during the first session displayed an opposite pattern during the second exposure 15 days later. This study suggests that well before the development of language, there may be a form of early memory encoded in our biology through stress regulation patterns, specifically related to firsthand socio-emotional experiences. Even if this learning does not manifest behaviorally, it is conceivable that the modifications or adaptations observed at the neuroendocrine level implicitly contribute to the child's process of creating meaning in the relational world from the early stages of development.

The presence of embodied memories, embedded in biological and physiological regulatory profiles, holds significant importance from a scientific standpoint. This challenges the conventional understanding of memory, suggesting the existence of event memories before the emergence of language and the autobiographical self. This has profound implications for clinical practice, as any alterations or dysregulation of the HPA axis can be viewed as adaptive attempts rather than mere consequences of adverse circumstances. Understanding a child's developmental history can be immensely valuable for child healthcare professionals—educators, psychologists, teachers—enabling them to interpret children's evolving behavior in the context of hidden regulators that exert substantial influence on stress response, emotional regulation, and learning.

6   Genes and what you do

At the heart of various theories concerning human development is the notion that early life experiences significantly contribute to shaping the distinct individual differences observed in adult behavior. Chapter 3 of this volume explores some of these theoretical perspectives, and overall, the book consistently emphasizes the acknowledgment that each child is inherently tied to their caregiving context. However, substantial variations exist among individuals in terms of how they are affected, both in the immediate and prolonged periods, by life experiences and environmental influences.

The identification of the DNA molecule, constituting the essence of genes, has broadened our understanding of individual variations beyond mere behavioral observations. Despite the extensive and intricate path leading to this discovery, Watson and Crick are universally recognized for their association with the double helix, immortalized in a renowned photograph featuring them alongside their initial three-dimensional model. Preceding them, Miescher, working in Germany in 1871, had already isolated a novel organic compound named nuclein in fish sperm. Approximately two decades later, German scientists Kossel and Neumann demonstrated that this nucleic acid, abundant in phosphorus, encompassed four distinct types of bases, played a role in cytoplasmic synthesis, and it was a component of chromatin. In the 1930s, X-rays unveiled the helical structure of DNA, while approximately twenty years later it was demonstrated the consistent pairing of adenine with thymine, and cytosine with guanine, providing the foundation for Watson and Crick's model. In 1952, a year prior to their Nobel Prize accolades, the iconic "photo 51," an X-ray image captured by Rosalind Franklin, unequivocally depicted the spiral structure of DNA. A year later, Watson and Crick's concise initial publication, featuring the first illustration of the DNA double helix illustrated by Crick's wife, Odile Speed —was followed by a second paper in 1953 in the journal *Nature*. This subsequent publication provided more comprehensive details about their model, marking an irreversible turning point in the history of biology. Their ingenious model encapsulated all four fundamental principles essential for genetic material. First, it elucidated the replication of genetic material crucial for reproduction and life. Second, it clarified the specificity of genetic material, addressing the quality of genes and their preservation during duplication processes.

DOI: 10.4324/9781003479314-8

Third, the model expounded on the information content of genetic material, defining DNA as a molecular library. Last, it shed light on the adaptability of genetic material, showcasing its capacity to undergo mutation and change over time.

This discovery is relatively recent, especially considering how rapidly DNA has evolved into an iconic concept in popular culture, known not just to experts. The renowned "double helix" has become the symbol of biology and a pop icon in the contemporary era. The historian of science, Martin Kemp, went so far as to define the DNA molecule as the Mona Lisa of modern science (Kemp, 2003). Since the revelation of the double helix, which unveiled the molecular secret of life more than 60 years ago, the collective imagination has been actively constructing a parallel library, enabling individuals to interpret and sequence their own meanings of what DNA represents. In popular culture, the DNA molecule has spawned a diverse array of artistic expressions. For instance, in 1963, Salvador Dali painted a work in homage to Crick and Watson entitled "Galacidalacidesoxiribunucleicacid." In an interview with Playboy magazine, Dali even declared that the announcement of the discovery of DNA was definitive proof of the existence of God for him (Smith, 2016). In the same year, Stan Lee and Jack Kirby created the X-Men, a group of troubled and outcast superheroes where the X not only referred to Professor Xavier but also denoted their mutant nature—the X chromosome, a note. In the 1970s, David Bowie referred to three of his albums, collectively known as the "Berlin trilogy"—Low, Heroes, and Lodger—as his DNA. The significance of DNA is apparent in various films and TV shows. For instance, in the 1982 film "E.T. the Extraterrestrial," when scientists celebrated the discovery of its DNA, viewers grasped the importance of this revelation. From the extraction of DNA from a mosquito encased in fossil amber, enabling John Hammond to realize the dream of Jurassic Park in 1993, to the genetic tampering of Dana Scully's DNA in the second season of "The X-Files" in 1994, DNA has emerged as a central theme in storytelling. Whether it's the clone wars in the Star Wars universe (2002) or the success of forensic genetics on television, as seen in TV series like "C.S.I." (aired from 2000 to 2015), DNA has continued to be a prominent pop icon to this day.

Genetics and psychology, a difficult relationship?

In the early 1980s, the integration between the fields of behavioral genetics and developmental psychology was in its early stages. Professor Robert Plomin, hailing from the University of Colorado, highlighted in 1983 that this collaboration had been slow and gradual, characterized by what he termed a "certain torpor." One factor contributing to this hesitancy could be traced back to the challenge of relinquishing a specific notion of genetics, where the influence of DNA was thought to be confined to the period before birth, maintaining a fixed impact throughout one's life. This apparent resistance of genetics to acknowledge a developmental dimension served as a rationale for the somewhat tepid enthusiasm among scholars in psychology and behavior from an evolutionary standpoint. Both sides, however, played a role in fostering this misconception. Behavioral genetics proposed the existence of genes with structural constancy, portraying them as constants rather than dynamic factors that accompany and modify development. The conception of an unchanging and immutable genetic model has been disrupted by the emergence of new genetics, including epigenetics, which has revealed how the regulation of DNA functioning is significantly influenced by the environment. It also demonstrated how slight genetic variations can lead to diverse and complex arrangements of phenotype and endophenotype.

When behavioral genetics methods are applied to human development, they can investigate the sources of genetic regulation and elements contributing to continuity and stability throughout the developmental process. Another factor complicating the relationship between genetics and developmental psychology is a prevailing normative perspective on human development, which primarily aims to describe an average and generalized pattern of development. This approach has long dominated developmental psychology, with the study of individual variations only recently gaining prominence through the neo-constructivist paradigm. Normative inquiries are concerned with the typical development of individuals, whereas behavioral genetics directs its focus towards understanding individual differences and their underlying causes.

Psychologists' hesitancy in accepting the role of behavioral genetics may also stem from reservations about its clinical implications. Adjusting the conditions of care through environmental interventions is a more conceivable prospect compared to the seemingly ambitious task of modifying an individual's genome. Nonetheless, it is crucial to underscore that clinical developmental psychology stands to gain valuable insights from research that documents individual responses to environmental stimuli and early interventions. Psychology can play a pivotal role in steering the scientific advancements in behavioral genetics toward practicality and adaptability within a clinical context. A comprehensive understanding of human genetic variation has the potential to facilitate the development and implementation of effective environmental interventions, thereby enhancing the quality of life for individuals.

In conclusion, according to psychologists Scarr and Weinberg (1980), the essential factors for an individual's growth and development encompass both genes and the environment. Behavioral genetics, an interdisciplinary science drawing on insights and methodologies from fields such as psychology, molecular genetics, and neuroscience, is dedicated to exploring the intricate relationships between genes and the environment. It seeks to comprehend the contributions of both factors to an individual's evolutionary trajectories. Quantitative genetics involves the comparison of diverse family members, including twins or adopted children, whereas molecular genetics concentrates on specific genetic markers in DNA, which may play a role in identifying individual variations and differences (Plomin et al., 2016).

From quantitative genetics to molecular genetics

The quantitative genetics approach heavily relies on investigations involving monozygotic or dizygotic twins. Twins serve as an ideal natural experiment, presenting a scenario where two individuals share 100% of their genetic heritage if they are homozygous, and approximately 50% if they are dizygotic. Growing up to-gether, twins not only share their genetic makeup but also a significant portion of the environmental context, encompassing aspects such as parents, socioeconomic conditions, and often, school, class, and teachers (Plomin et al., 2016). The study of twins provides an avenue to explore the distinct contributions of shared environ-ments, which may account for similarities between twins, as well as non-shared or unique environments, offering insights into individual differences. The central question arises: "How do children raised in similar contexts exhibit notable differ-ences?" However, seemingly shared elements in the environment may only be par-tially alike in the twins' experiences (Plomin et al., 2013). Variables such as a low socioeconomic level could motivate one child to strive for a particular outcome while prompting another child to abandon efforts prematurely. These "subjective" intersections at the interface between the individual and the environment may be influenced by a multitude of genetic and environmental factors.

The twin method facilitates the estimation of the heritability of specific behav-ioral traits or characteristics, indicating the role of genetic factors in individual variations. The heritability can be calculated as twice the difference between the correlations of a particular trait in monozygotes and dizygotes. For instance, if the correlations among monozygotes for mathematical ability are 0.68 and for dizy-gotes it is 0.41, the heritability is estimated to be 0.54 (twice 0.68 minus 0.41). This implies that 54% of the variability in that specific ability can be attributed to the sharing of genetic heritage. The impact of the shared environment is de-termined by subtracting the correlation for monozygotes (0.68 in this example) from the heritability value (0.54). In this hypothetical scenario, about 14% of the variability would be accounted for by the shared environment. The influence of the non-shared or unique environment is derived by subtracting the correlation value for monozygotes (0.68) from the total variability (1.00). In our example, the non-shared environment would elucidate approximately 32% of the variability in the targeted mathematical skills.

Twin registries serve as invaluable repositories for investigating the influences of genetics and the environment on intricate psychological and behavioral charac-teristics. These registries encourage collaboration across disciplines and streamline the exploration of gene-environment interactions. Notable examples in Europe in-clude the Danish Twin Registry, boasting over 85,000 twin pairs (Skytthe et al., 2013), the German Twin Study on Cognitive Ability, Self-Reported Motivation, and School Achievement (CoSMoS) (Hahn et al., 2013), and the Young Nether-lands Twin Register (YNTR), encompassing more than 70,000 twin pairs born in the Netherlands since 1985 (van Beijsterveldt et al., 2013). These registries aspire to offer a comprehensive depiction of twins' phenotypes and endophenotypes, cov-ering variables related to health, disease status, and neurophysiological attributes.

Recent advancements in our comprehension of the intricate mechanisms gov-erning DNA and genes have made significant strides. Each individual possesses a distinctive genetic profile that plays a role, to varying extents, in shaping and mani-festing their phenotype throughout development. The human genome comprises 3 billion base pairs of nucleotides, with a mere 2% constituting genes. Less than 1%

of the sequence exhibits variation among different individuals, contributing to the noteworthy diversity of phenotypes within a species. These variations often involve polymorphisms, encompassing alterations in the sequence of bases such as modifications to a nitrogenous base, replacements, insertions, or deletions in portions of the genetic "text" (Plomin et al., 2013). Research associating these structural or sequential changes with specific behavioral phenotypes reveals that common clinical conditions or developmental disorders arise from an array of genetic factors, each contributing to a small fraction of the overall effect. These factors are commonly referred to as quantitative trait loci (QTL). When a trait is influenced by multiple genetic factors, it is distributed quantitatively in the population, and developmental disorders can be perceived as one end of a spectrum, with QTLs viewed as probabilistic elements rather than deterministic components.

The identification of genetic factors involves utilizing DNA markers, such as single nucleotide polymorphisms (SNPs), which are alterations of a single nucleotide in the DNA sequence. While individual SNPs may exert only a modest impact on observed behavior, aggregating multiple polymorphisms into a polygenic index allows for the determination of more substantial portions of variance in the role played by genetic variations in child development. Genome-wide association studies (GWAS) have marked significant progress in molecular genetics. These studies simultaneously assess hundreds of thousands of genetic markers to explore their association with specific outcomes. The utilization of DNA chips or microarrays containing millions of distinct segments enables more cost-effective and rapid genotyping. Analyzing large samples comprising hundreds of participants facilitates the identification of numerous genetic markers. By 2010, GWAS had established associations between over 700 genetic variants and more than 100 specific behavioral traits, while also revealing new polymorphisms not previously explored. However, the replication of individual associations necessitates further investigation through independent samples (McCarthy et al., 2008).

Vulnerable to stress

One of the most fruitful domains for the integration of developmental psychology and behavioral genetics involves exploring the reciprocal influences between genes and the environment in shaping an individual's response to stressful contexts. A prominent model for comprehending individual differences in reactions to stressful situations is the diathesis-stress model (Heim & Nemeroff, 1999). This model suggests that certain individuals, owing to inherent vulnerabilities, may be more or less impacted by exposure to stressful experiences. These vulnerabilities can manifest at various levels, encompassing behavioral, physiological, constitutional, or genetic factors. Stressful experiences may include neglect, maltreatment, acute traumatic events, as well as less severe circumstances like low parental sensitivity or chronic unfavorable conditions over time. According to the diathesis-stress model, a child with a challenging temperament or carrying a genetic vulnerability may display maladaptive functioning in response to unfavorable environmental events, potentially leading to psychopathology later in life. The model implies a condition of double risk, involving constitutional vulnerability and exposure to adverse environmental circumstances.

While this model is supported by numerous empirical findings, subsequent studies have indicated that it offers only a partial perspective on the complex processes and mechanisms involved in the interplay between nature (genetics) and nurture (caregiving) (Roisman et al., 2012). The success of this model is partly influenced by implicit assumptions of behavioral scholars. Notably, the scientific literature in clinical developmental psychology has primarily concentrated on examining the impact of early adverse events and unfavorable environmental conditions, aiming to comprehend how these factors contribute to poorly adaptive developmental outcomes and how they can be prevented. In contrast, less attention has been directed towards studying the effects of early protective events, favorable environmental conditions, and optimal experiences (Belsky & Pluess, 2009).

The expansion of this model is attributed to the independent contributions of Jay Belsky (1997) and Boyce and Ellis (2005). They have presented distinct arguments—theoretical-evolutionary by Belsky and psychobiological by Boyce and Ellis—challenging the notion of genetic vulnerability in favor of a perspective rooted in plasticity and evolutionary adaptation (Ellis et al., 2011). Despite differing assumptions, as discussed in the subsequent paragraphs, both scholars have contributed to a paradigm shift in the current understanding of the interaction between genetics and the environment. Importantly, they have introduced the widely accepted notion that factors once considered vulnerability markers—such as a specific genotype or temperamental trait—are, in reality, parameters of plasticity that confer greater or lesser flexibility and adaptation to an individual in response to both negative and positive environmental exposures (Belsky et al., 2007; Belsky & Pluess, 2009).

Evolutionary reasons

The fundamental question of why early experiences exert a significant influence on development underpins Belsky's hypothesis. In the early 1990s, Belsky and colleagues' (1991) research on the evolutionary origins of socialization faced criticism from advocates of behavioral genetics. Belsky's socialization theory proposed that both stressful and protective environments shape family dynamics, particularly parent-child and parent-parent relationships, ultimately impacting the socioemotional and behavioral development of the child and, subsequently, the future adaptation of the adult. Conversely, behavioral genetics studies suggested that variations in child development were primarily attributable to shared genes between parents and child, offering a genetic mediation explanation.

These criticisms led Belsky to ponder why early experiences should carry such significance in the evolution of our species, especially in human development. From a biological and evolutionary standpoint, the pivotal role of early experiences would only make sense if there was a reasonable certainty that future living conditions would resemble those of early experiences, at least within a single generation. However, the future remains largely unpredictable and uncertain. Consequently, the hypothesis that a single evolutionary mechanism exists, capable of utilizing early experiences to programmatically shape an individual's adaptation to future environmental conditions, appears problematic.

Moreover, from the perspective of modern evolutionary biology, it is challenging to conceive that natural selection acts solely on individuals for their survival; rather, it is more plausible that it operates to maximize reproductive success. Reproductive fitness involves transmitting genes to subsequent generations, while inclusive fitness considers indirect reproductive success, which can occur through other individuals sharing the genetic heritage of a particular individual. As the future is inherently unknown, parents have always been unaware of the optimal practices ensuring the reproductive success of their offspring and, consequently, their own indirect reproductive success in terms of inclusive fitness. Therefore, following evolutionary adaptation strategies to enhance reproductive success, natural selection would influence parents to raise children varying in neurodevelopmental plasticity (Belsky & Pluess, 2009). Consequently, if a parenting practice proves counterproductive in terms of evolutionary adaptation, "resistant" children, less affected by changes in parental behavior, would not manifest the behavioral consequences of this disadvantageous trait and would not pass it on to future generations. From an inclusive fitness perspective, the reduced susceptibility of "resistant" or less malleable children would result not only in an immediate intragenerational advantage but also an indirect benefit for subsequent generations. Simultaneously, parenting practices promoting better adaptation would benefit highly plastic or malleable individuals and their descendants.

This theoretical reasoning on evolutionary foundations leads directly to the hypothesis that children vary in their degree of plasticity and susceptibility to parental care practices and environmental influences more generally. In addition, while Belsky's proposal does not negate the possibility of further environmental influences on susceptibility gradients, the differential susceptibility hypothesis, at least in its initial formulation, primarily rests on the role played by nature and genetic variations in creating opportunities for flexibility and adaptation that maximize an individual's reproductive success, whether directly or indirectly. Empirical evidence supporting Belsky's approach, its alignment with empirical data, can be found in research concerning animal models. For example, Nussey and colleagues (2005) demonstrated that between the 1970s and early 2000s, natural selection appeared to favor individuals with high plasticity regarding the timing of reproduction, likely in response to pressures from ongoing climate change, resulting in a mismatch between the timing of care in birds and their predators. Suomi (2006) argued that humans and macaques (rhesus monkeys) exhibit particularly impressive resilience compared to other primate species. Both humans and rhesus monkeys can survive and adapt in vastly different environmental contexts, geologically and socially, and they are highly likely to thrive and flourish when introduced to new environments. Thus, it is conceivable that one of the factors contributing to the relative adaptive success of humans and macaques lies not so much in genetic specializations but rather in a more general genetic variation that results in a high degree of plasticity and malleability. In other words, genetic diversity may hold one of the secrets to resilience and adaptive ability.

Psychobiological reasons

In 1995, Bruce Ellis and colleagues published a paper in *Psychosomatic Medicine* that offered additional validation for the theoretical propositions of differential susceptibility. The study underscored the significance of psychobiological and psychophysiological processes in the interplay between genetic and phenotypic variations, presenting a measurable and investigable mechanism to complement Belsky's psychoevolutionary conjectures. The research investigated the interactive influences of changes in exposure to environmental stressors and constitutional biological reactivity in relation to the risk of developing respiratory disease. One study assessed biological reactivity at the autonomic nervous system level in response to stress in the laboratory, while the second study evaluated the immune response to stress associated with entering school.

The findings indicated that children with lower cardiovascular or immune reactivity to stress showed no significant differences in the likelihood of developing respiratory disease, regardless of whether they resided in high- or low-risk settings. In line with the prevailing diathesis-stress model, children with high physiological reactivity to stress exposed to high-risk environmental conditions exhibited a higher incidence of respiratory diseases compared to other children. However, the most unexpected discovery was that children with a high stress reactivity profile, living in low-risk conditions, displayed the lowest rate of respiratory disease compared to all other groups of children. The authors speculated that some form of "biological sensitivity to the characteristics of the social world" was at play (Boyce et al., 1995).

Expanding on these investigations and subsequent scientific findings, Boyce and Ellis (2005) formulated a theoretical model of biological sensitivity to the environment. This model delineates specific psychobiological mechanisms related to stress reactivity that (1) have evolved to address particular challenges linked to social competition, survival, and reproduction throughout evolutionary history; (2) can respond adaptively to relatively stable conditions of protection and support or to environmental stress and adversity; and (3) are activated in a specific manner based on different thresholds of responsiveness to the environment, enabling the organism to achieve optimal adaptation.

In this context, "adaptation" denotes the regulation of physiological and biological systems that enhance the likelihood of evolutionary success, encompassing aspects such as health, growth, survival, and reproduction. Boyce and Ellis also posit that this regulation operates "for better or worse," responding to both stressful and protective environmental stimuli. Consequently, children may vary not only in their resistance or vulnerability to adverse experiences but also in their capacity to adapt and be flexible in response to diverse environmental exposures, whether in risk or protective contexts.

Building on this framework, children with a relatively robust ability to withstand particularly adverse environmental situations were labeled "dandelion" children, while those exhibiting high sensitivity and physiological reactivity to both adverse stressful and positive-protective situations were termed "orchid" children. These terms are derived from Swedish idiomatic expressions, "maskros-barn" (dandelion child) and "orkide-barn" (orchid child), used to characterize children who can manage well even in the worst situations and children who are highly sensitive to external stimulations, respectively.

Comparable constitutional variations in sensitivity to protective or risky social contexts have been observed in juvenile macaques by Suomi (Barr et al., 2004). In highly stressful environments, the most biologically reactive and sensitive macaques, termed "orchid" individuals, were more prone to display maladaptive developmental outcomes. However, in protective environments, these macaques exhibited superior evolutionary success compared to their counterparts who were relatively less sensitive and responsive to the environment, known as "dandelion" individuals.

Furthermore, Boyce and Ellis put forth a curved U-shaped relationship to elucidate the link between environmental stress levels in early childhood experiences and the development of a profile characterized by disproportionate stress reactivity. In particularly supportive contexts, it would be adaptive for children to be highly influenced by the enriching and optimal stimuli of their early care environment. This heightened influence could enhance their social competitiveness and adaptive abilities, ultimately ensuring reproductive success. Conversely, growing up in perilous or disadvantaged conditions would be advantageous for survival and adaptive success if the child developed heightened vigilance to threats, preparing them to cope with similarly disadvantageous future environments.

Better than vulnerable: Malleable

The work of Boyce and colleagues (1995) and Belsky (1997) was actually fore-shadowed—at least 15 years earlier—by the contributions of Wachs and Gandour (1983), who had already advocated for using the concept of "organism specificity" to underscore similar variations in the effects of the environment on phenotype development. In their paper, they proposed that the observation of differential reactivity in different individuals in response to similar environmental stimuli should prompt a focus on the study of individual variations as a central aspect of research at the intersection of social and biological sciences. This exemplifies how science often advances through unexpected convergences of independent contributions until the conditions for a paradigm shift emerge.

The core thesis of the proposals of differential susceptibility (Belsky & Pluess, 2009) and biological sensitivity to context (Boyce and Ellis, 2005) is to argue that those who are generally identified according to the diathesis-stress model as "vulnerable individuals"—being more influenced than others by adverse experiences—are also the same individuals who, when exposed to supportive and protective environments, emerge strengthened to the extent of displaying more adaptive functioning than so-called resistant individuals. In other words, there would be individual differences in plasticity and susceptibility to environmental influences, where some individuals would be largely susceptible to environmental exposures—for better or worse—while others would be more resilient. This view contrasts with the traditional diathesis-stress model, which considered individuals vulnerable to adverse exposures but did not predict variations in response to exposure to protective and favorable environmental conditions. Unlike Boyce's model, the diathesis-stress model cannot predict that the attributes and mechanisms making an individual susceptible to developing maladjustments in response to unfavorable care contexts would also increase the benefits for the same subject in protective and caring environments.

Although Belsky's proposal and that of Boyce and Ellis agree on many points, they also have significant differences. First, Belsky developed the theoretical framework of the differential susceptibility hypothesis based on speculative grounds, not from empirical data, and subsequently sought evidence in the scientific literature to support this hypothesis. Therefore, identifying factors or mechanisms responsible for the different susceptibility gradients is not of primary interest to Belsky. On the contrary, Boyce's thesis of biological sensitivity to context considers the identification of these mechanisms as central to the assumptions of the theoretical model. In addition, according to the theory of biological sensitivity to context, different physiological or neurobiological mechanisms underlie different reactivity profiles, while Belsky refers mostly to behavioral and temperamental reactivity factors. This is not a minor point, considering that attempts to find overlap between behavioral and physiological-biological reactivity have often failed in both adults and children. Moreover, Boyce and Ellis emphasize the role played by the interaction between genes and the environment, while Belsky focuses on heritable variations in susceptibility to care experiences without specifying the environmental antecedents of these variations (Boyce & Ellis, 2005). Finally, while the model of biological sensitivity to context suggests that orchid children may have particular difficulty adapting to adverse contexts but are facilitated in socially and emotionally supportive contexts, Belsky conceptualizes emotionally temperamentally responsive children as more malleable and plastic, capable of best adapting to a wide spectrum of environmental conditions (Boyce & Ellis, 2005).

Conditional adjustments

In the past three decades, scholars and researchers in evolutionary biology have come to acknowledge that evolution did not likely prescribe a singular, universally optimal strategy for survival and reproductive success across species. This is because the effectiveness of a strategy depends on various unpredictable factors like physical, economic, social, and cultural conditions, making it challenging to predict the best approach. In essence, a strategy that enhances evolutionary success in one environment may prove inadequate in others. Consequently, the pressures of natural selection would favor an adaptive plasticity of the phenotype, enabling a single genotype to generate a diverse array of phenotypes in response to specific ecological conditions that have recurrently occurred throughout the evolutionary history of a species (Belsky et al., 1991). The development of alternative phenotypes is not a random occurrence; rather, it stems from iterative interactions between genes and the environment, fostering an enhanced ability and inclination of individuals to modify their phenotype in response to environmental variations during development.

The early developmental environment holds particular significance in this context. During this period, numerous complex adaptations occur, and it is advantageous to establish a system capable of responding reliably and non-randomly to stress through specific physiological and neuroendocrine strategies as early as possible and in the most flexible manner conceivable (Davidson et al., 2000). In addition, certain behavioral strategies undergo refinement in the early years through repeated exposure and experience (Draper & Harpending, 1982). Nevertheless, an excess of plasticity or malleability in response to the environment should not endure in later developmental stages. A dynamic system must evolve to establish attractive states defined by specific threshold values, ensuring stability and continuity in space and time. Simultaneously, the system must retain the ability to provide responses at the right time and contingent upon the current context. Adopting a strategy where one regulates behavior solely based on learned or preprogrammed strategies that lack room for further revision would be disadvantageous, especially in competitive or social cooperation scenarios.

One can propose a hierarchy of mechanisms that facilitates learning from experience and offers contingent responses sensitive to the environment. At the apex of this hierarchy, psychological mechanisms of a socio-cognitive and socio-emotional nature would be situated, capable of providing the environmental assessment and interpretation necessary for flexible behavior. At lower hierarchical levels, there would be anatomical, physiological, endocrinological systems, and so forth, endowed with the ability to store prior experiences with the environment in profiles of reactivity and biological regulation—a position akin to that expounded in Chapter 4 concerning the evolutionary origins of the relationship between the autonomic nervous system and social cognition (Porges, 2001). In concert, this hierarchy would ensure the feasibility of conditional adaptations reflecting the interplay between genes and the environment, fostering the development of flexible strategies, and facilitating the adaptive response of the individual to diverse contexts.

An instance of such conditional adaptations is evident in the animal kingdom. Take for example *Nemoria arizonaria*, a caterpillar that develops entirely distinct morphologies based on its diet during the initial three days of life. This species inhabits oak forests in southwestern America during spring and summer. Despite both broods having a similar appearance at different times of the year, the spring brood, which feeds on oak inflorescences, adopts an appearance akin to that of a flower, while the summer brood, which feeds on oak leaves, exhibits a morphology more reminiscent of twigs. Consequently, *N. arizonaria* can successfully camouflage itself during various seasons. In humans, supporting evidence for these conditional adaptations—developmental outcomes that affirm the efficacy of the differential susceptibility or biological sensitivity to context model—can be observed both at the behavioral and genetic levels. The following two paragraphs provide some insights, although a comprehensive treatment would likely necessitate a dedicated volume.

Temperament and sensitivity

Behavioral investigations into temperament and emotional regulation in very young children offer compelling evidence for differential susceptibility. Even before the advent of behavioral genetics studies, often framed as Genes × Environment research, the influence of parental behavior on temperament profiles and regulation in the initial months of life had been explored in several studies (Belsky, 2005). Numerous investigations have demonstrated how various components of parenting practices—such as discipline, interactive synchrony, and maternal sensitivity—account for portions of the variance in a child's behavioral outcomes across different domains, encompassing self-regulation skills, externalizing problems, inhibition, and irritability (Belsky et al., 1998; Feldman et al., 1999; Kochanska, 1993). In essence, fluctuations in parental behavior and the quality of early care contribute to variability in child developmental outcomes. Some studies have revealed these effects not only in terms of vulnerability under diathesis-stress but also in terms of the influences of differential susceptibility, capable of shaping the plasticity of a child's development for better or worse. For instance, van Aken and colleagues (2007) found that 16- to 19-month-old children with a challenging temperament exhibited fewer externalizing problems when raised by highly sensitive mothers, while more frequent externalizing problems were observed when mothers employed controlling strategies and insensitive responses. This gradient of plasticity was not evident in children without a challenging temperament. Large-scale studies in the United States revealed how the quality of parental care, quantitatively measured through multiple play sessions in the first year of life, negatively predicted a child's behavioral problems at school entry, with this effect moderated by the child's temperament. Specifically, in children characterized by a challenging temperament, higher-quality parenting was associated with fewer behavioral problems, whereas lower ratings in the quality of early care predicted greater behavioral problems (Bradley & Corwyn, 2008). Similar to studies on challenging temperament, other reactivity traits to environmental stimuli have also been identified as factors of differential susceptibility. For instance, children aged 15–28 months displaying a greater inclination toward angry responses, compared to less reactive children, exhibited fewer behavioral problems at five years of age when raised by parents who enforced clear rules. Conversely, those with less consistent and clear discipline from their parents manifested more behavioral problems.

Genetic variations

At this juncture, it becomes apparent that both the theory of differential susceptibility and the concept of biological sensitivity to context delineate individual disparities in evolutionary plasticity as the pivotal factor—a sort of sliding doors—that can elucidate not only the impact of adverse experiences on maladaptive developmental outcomes but also the role of supportive and favorable environments in fostering evolutionary accomplishments, resilience, and stress regulation. As mentioned, this perspective starkly contrasts with the classical diathesis-stress model, which underpins the majority of initial studies in behavioral genetics—Gene × Environment (G × E) studies, to be precise. These studies traditionally sought to identify certain traits indicative of vulnerability in individuals more prone than others to endure the negative consequences of adverse developmental contexts, particularly in early childhood (Monroe and Simons, 1991; Zuckerman, 1999).

Specific polymorphisms have been extensively scrutinized concerning their role in vulnerability, as per the diathesis-stress model, or plasticity, as per the perspective advocated by Belsky, in genetically mediating the impact of environmental conditions on the socio-emotional and cognitive development of the child. The MAOA gene, located on the X chromosome, encodes the MAOA enzyme responsible for metabolizing neurotransmitters—norepinephrine, serotonin, dopamine—rendering them inactive. The correlation between a genetic variant of MAOA and early experiences of abuse and neglect contributing to antisocial behavior prompted Caspi's research group (2002) to posit that genotype variations mediated the effects of early traumatic experiences. Studies conducted by Caspi and colleagues in New Zealand validated that among young males studied longitudinally, those with a low-activity form of the MAOA gene were more prone to adopt violent behaviors if exposed to maltreatment during childhood. While these findings were generally interpreted from a diathesis-stress standpoint, some researchers hypothesized that those most susceptible to the adverse effects of maltreatment were also those exhibiting lower levels of antisocial behavior in the absence of such experiences in early life. Notably, seven-year-old children possessing the low-activity variant of the MAOA gene had a heightened risk of psychopathology if subjected to abuse but fewer mental problems—particularly attentional deficits—if not exposed to traumatic experiences, compared to age-matched individuals with the high-activity variant of the same gene (Kim-Cohen et al., 2006). Subsequent replications of these results in substantial cohorts of children and adolescents (Enoch et al., 2010; Frazzetto et al., 2007; Widom and Brzustowicz, 2006) suggest that genetic variations in the polymorphism regulating MAOA expression confer greater plasticity, rather than heightened vulnerability.

Another extensively studied polymorphism in developmental behavioral genetics involves the serotonergic system, specifically the polymorphic region linked to the gene encoding the serotonin transporter linked polymorphic region (5-HTTLPR). Despite the documentation of several variants, the focus of most studies has been on two allelic variants: a short (S) and a long (L) variant (Nakamura et al., 2000). The S variant is generally associated with reduced expression of the serotonin transporter, a pivotal molecule regulating serotonergic reabsorption in the inter-synaptic space. This transporter is considered vital in emotion, mood regulation, and the risk of developing depressive pathology (Canli & Lesch, 2007). Caspi and colleagues (2003) conducted seminal work on the role of this polymorphism in mediating the impact of early traumatic experiences on the risk of developing depressive symptoms and engaging in suicidal acts in adulthood (Taylor et al., 2006). Regarding the development of stress regulation abilities in young children, the 5-HTTLPR polymorphism has been linked to negative emotionality responses in various experimental settings. Parents described 2-month-old children homozygous for the S variant as generally more negative, although by 12 months, these same subjects exhibited less fear during a stressful laboratory procedure (Auerbach et al., 1999). At 4 months, children carrying at least one S allele responded with increased emotional negativity to specific caregiving activities compared to those homozygous for the L allele (Pauli-Pott et al., 2009). Lakatos and colleagues reported that 12-month-old children who were heterozygous or homozygous for the S allele of 5-HTTLPR showed greater resistance and anxiety responses in interaction with an adult stranger compared to children homozygous for the L allele. G × E studies also demonstrated an interaction effect of maternal anxiety and the 5-HTTLPR genotype on the degree of negative emotionality in children at 6 months of age (Pluess et al., 2011). Another study revealed that negative emotionality induced by stressful procedures could be moderated by a G × E interaction at the age of 12 months (Pauli-Pott et al., 2009). Furthermore, maternal behavior at 7 months and the 5-HTTLPR genotype significantly correlated with the development of a secure attachment style at 15 months of age (Barry et al., 2008). Intriguingly, in this study, children with at least one L-allele, despite being at a higher risk of developing emotional dysregulation, demonstrated better adaptability to maternal supportive behavior than their peers with the L-allele, confirming the plausibility of the differential susceptibility model. At 4 months, full-term infants exposed to the still-face procedure exhibited greater negative emotionality during still-face and reunion episodes in the presence of at least one S allele compared to children with an LL genotype (Montirosso et al., 2015). Nevertheless, the results also indicated that during both the still-face and reunion phases, LL infants did not exhibit an association between sensitive maternal behavior and levels of negative emotionality. Conversely, S-carrier children displayed a significant and inverse relationship between maternal sensitivity and negative emotionality during both episodes. In essence, the S-carrier subjects, manifesting greater emotional negativity, reported lower stress levels than the LL group if their mother displayed high levels of maternal sensitivity. Taken collectively, these studies suggest that the interaction between the quality of maternal behavior and the presence of a distinct genotype for the 5-HTTLPR polymorphism may contribute to favoring the adaptive development of the child's socio-emotional skills. Specifically, this interaction appears to represent a form of differential susceptibility, wherein different degrees of malleability to the protective role of maternal sensitivity are contingent on the specific genotype—LL or S-carrier—of the 5-HTTLPR polymorphism.

The dopaminergic system also presents evidence of G × E interactions aligning with the Belsky model and the tenets of the theoretical proposition of biological sensitivity to context. The central nervous system's regulation by dopamine influences attentional, motivational, and reward systems. There is particular interest in the gene polymorphism (DRD4) responsible for synthesizing D4 receptors for dopamine. This polymorphism's allelic variants differ in the number of repeats of base pairs in exon III. The variant with seven repeats has been identified as a factor contributing to vulnerability to ADHD (Faraone et al., 2001), impulsive behaviors (Kluger et al., 2002), and lower efficiency in dopamine reception in the brain (Robbins & Everitt, 1999). Individuals carrying this genetic variation are not only more susceptible to adverse environmental conditions but also appear to derive more benefits from exposure to protective and favorable early care conditions (van IJzendoorn and Bakermans-Kranenburg, 2006). In a longitudinal study, maternal sensitivity was assessed when the children were 10 months old and was significantly associated with behavioral problems reported more than two years later, but only in children exhibiting the seven-repeat version of DRD4. These children displayed a high level of externalizing behavioral problems if their mothers demonstrated poorly sensitive and responsive behavior, yet reported the lowest levels of externalizing behavior when their mothers were rated as highly sensitive. In a study evaluating the efficacy of a video-feedback intervention aimed at supporting parenting skills, the same Dutch group, led by van IJzendoorn, later demonstrated that the intervention was more effective in reducing problematic behaviors in children with the seven-repeat variant of the DRD4 gene (Bakermans-Kranenburg et al., 2008). These children, evaluated at the intervention's conclusion, exhibited lower neuroendocrine reactivity of the HPA axis compared to the same-age group without the same genotype if assigned to the group of mothers receiving the video-feedback intervention. Conversely, those assigned to the control group demonstrated significantly greater cortisol production than their counterparts without the seven-repeat variant of DRD4.

Another polymorphism linked to individual differences in emotional regulation in children involves the gene responsible for encoding an enzyme that catabolizes catecholamines (COMT), thereby acting as a modulator of the sympathetic and adrenocortical response to stress (Glaser et al., 2005). The relevant polymorphism, known as COMT val158met, is characterized by the substitution of methionine for valine at codon 158, resulting in a two- to fourfold decrease in COMT enzyme activity (Chen et al., 2004). Previous findings have indicated that the variant with lower enzyme activity, the met allele, is associated with an increased vulnerability to developing affective disorders in adulthood (Aberg et al., 2011; Olsson et al., 2007). Furthermore, variations in the COMT genotype lead to distinct stress response profiles in adults. Met allele carriers, also referred to as met-carriers, exhibit a heightened stress response to painful stimulation (Zubieta et al., 2003) and increased salivary cortisol production in reaction to psychosocial stress (Jabbi et al., 2007). Although limited studies have reported such associations at an early age, the evidence appears to be inconsistent. When compared to individuals homozygous for the val variant, 9-month-old infants with at least one met allele demonstrate a reduced capacity for parasympathetic stress regulation (Mueller et al., 2014) and increased cortisol secretion in response to psychosocial stressful stimuli (Armbruster et al., 2012). In a more recent study, 4-month-old infants underwent the still-face procedure and were subsequently genotyped for COMTval158met through saliva samples. To obtain a quantifiable response of the HPA axis (Provenzi et al., 2016c), a modified version of the still-face paradigm was employed, maintaining the usual

A-B-A structure followed by a second still-face and a second reunion, resulting in a total duration of 10 min. The study revealed that met-carrier children exhibited greater negative emotionality during the second still-face compared to the equivalent age group with the val allele.

Positive and negative: Does it make sense?

This chapter has provided an introductory, albeit not exhaustive, exploration of some of the relationships between genetics and development. Non-Mendelian quantitative genetic research has shifted its emphasis from merely describing how genetic influences contribute to individual differences to comprehending how genes impact behavior. Approaches such as studying longitudinal cohorts throughout the life cycle and investigating genes and polymorphisms in connection with specific behavioral or endophenotypic outcomes exemplify this shift. The pivotal contribution of G × E studies in this domain has significantly influenced the discourse on the interactions and correlations between nature and culture, genetics and the environment, surpassing the Mendelian logic of heredity. Notably, the majority of the content in this chapter has focused on exploring the interactions between genetics and the environment concerning the early development of children in their initial years of life.

At present, it is widely acknowledged that the single-gene or gene-target approach falls short of providing comprehensive insights. The development of specific phenotypic traits is influenced by numerous genes, each contributing with modest effect sizes. Consequently, studies involving large samples are necessary to capture these subtle effects. Scholars in behavioral genetics continue to grapple with the challenge of "missing heritability" (Plomin, 2013). Rather than attempting to replicate studies focusing on individual genes or polymorphisms, which yield significant but small effect sizes, the preferred approach is for researchers at the intersection of developmental psychology and behavioral genetics to collaborate. Together, they can gather behavioral data and DNA from sufficiently large samples to uncover robust associations. In these extensive cohorts, the use of polygenetic composite indices is favored, allowing for the identification of genetic risk and resilience factors. This collaborative effort enhances our understanding of the intertwined roles of genetics and environment in human development.

In this chapter, the discussion has frequently touched upon the significance of unfavorable, adverse, optimal, and protective environments in shaping the development of individuals with varying susceptibilities or sensitivities to environmental changes. A subtle yet noteworthy aspect that may not have eluded the observant reader is the inquiry into what designates an environment as unfavorable, adverse, optimal, or protective. As previously noted, studies in developmental behavioral genetics have predominantly focused on exploring the repercussions of early traumatic experiences on the behavioral or neurobiological development of children. This emphasis on unfavorable environments has perpetuated a somewhat incomplete theoretical model. In these investigations, control groups often comprised individuals lacking the same constitutional deficits but exposed to a similar environmental disadvantage, or individuals sharing the same vulnerability but not subjected to equivalent traumatic or stressful experiences. Nevertheless, characterizing the absence of adversity simply as an optimal or "good" context may not be the most accurate approach. Such a perspective might lead to an underestimation of outcomes related to differential susceptibility and, conversely, an overestimation of extreme vulnerability.

Determining what qualifies as a good or bad environment, a favorable or unfavorable condition, presents a challenge due to the inherent complexity of a non-linear dynamic system, as discussed in the initial chapter of this book. The question itself carries an inherent bias, as the equilibrium of such a system does not culminate in a state of ultimate adaptation. Instead, development unfolds as an unpredictable sequence of dynamic equilibrium states, where adaptation manifests not as a fixed outcome but as a transient state influenced by specific environmental characteristics and system parameter oscillations within relatively close threshold values. Adaptation, in this context, emerges through mutual and reciprocal interaction between the individual and the environment, recognizing the individual as an integral part of the environment. The question of what constitutes a favorable or unfavorable environment runs the risk of being misleading, potentially prompting researchers to pursue an investigation aimed at categorizing certain environmental conditions as either positive or negative, albeit with partial and illusory certainty. The dichotomy of "positive" and "negative" appears overly simplistic to those with clinical experience. The designation of "positive parenting" in scientific literature, while not uncommon, poses risks as it tends to oversimplify the observation, lacking the necessary granularity to characterize what truly renders an environmental condition positive, protective, or supportive. Particularly in the clinical translation of behavioral genetics studies, the reference to positive or negative environments can be detrimental, reinforcing the misleading notion that certain environments or behaviors are universally protective. This assumption is potentially dangerous in light of the theoretical proposals and empirical evidence presented in this chapter. While behavioral genetics studies aid in identifying early risk profiles and proposing mechanisms and efficacy gradients for early interventions at the individual level, there remains a considerable level of uncertainty regarding what can happen in the caregiving environment. Cultures and parents exercise their prerogatives to determine what they consider feasible and necessary. The challenge for developmental psychobiology is to advance our understanding of how, within a dynamic non-linear system, the caregiving environment aligns its prerogatives harmoniously and coherently with the unique characteristics and sensitivity of newborns and children.

7   Footprints in the epigenome

Paris and London

At the onset of the nineteenth century, Paris bore the recent impact of the Bastille's storming and the execution of rulers Louis XVI and Marie Antoinette. Merely two years later, Napoleon Bonaparte would ascend to the imperial throne at Notre-Dame Cathedral. Preceding these events, a French marquis and chemist duo achieved the inaugural balloon flight using a creation by the Montgolfier brothers. Simultaneously, construction projects for landmarks like the Arc de Triomphe, Montmartre, and Père Lachaise commenced, while work on the Louvre persisted. Against this backdrop, in 1809, the French naturalist Jean-Baptiste de Lamarck, known simply as Lamarck, released his renowned work, "Zoological Philosophy." For the first time, he presented a theory on the adaptation of living systems grounded in evolution. Lamarck's treatise introduced the novel concept that the environment plays a crucial role in fostering phenotypic changes linked to phylogenetic shifts throughout evolution. He proposed two guiding principles for environmental influences: the law of use and non-use, wherein an organ develops with increased usage, and the inheritance of acquired characters, positing that all acquired traits would be passed on to the succeeding generation.

Half a century later, London stood as the world's most sprawling city, boasting the largest port and serving as the epicenter of international trade. The construction of the world's most iconic underground system was on the brink, while the British Museum and National Gallery had already become integral parts of the city's cultural legacy. Positioned as the capital of the British Empire, London attracted a convergence of numerous migratory streams, poised to transform into a genuinely cosmopolitan city. In 1851, the Crystal Palace in Hyde Park hosted the opening of the Great Exhibition. Coinciding with this momentous event, exactly 50 years after Lamarck's work was published, British biologist Charles Darwin released "The Origin of Species." In this groundbreaking work, Darwin proposed his own evolutionary theory, centered on the concept of random natural mutations and the environment's role as a selector favoring traits conducive to survival and censoring maladaptive ones. According to Darwin's perspective, not all acquired characteristics endure; survival is reserved exclusively for the fittest.

DOI: 10.4324/9781003479314-9

The rediscovery of Mendel's work at the outset of the twentieth century marked a pivotal moment, ushering in a gradual triumph for Darwin's theories, the emergence of neo-Darwinism, and the rejection of Lamarck's propositions. In the mid-twentieth century, Luria and Delbrück (1943) executed the fluctuation test, a notable experiment. A liquid culture of high-concentration bacteria was plated both in the absence and presence of phage until a plate was obtained wherein the majority of bacteria perished, yet small colonies of survivors emerged. This outcome could be interpreted through the lens of the Lamarckian hypothesis of adaptation, suggesting that resistance to contaminants could be acquired through direct exposure to the polluted environment. Alternatively, it could align with Darwin's mutational or random hypothesis, indicating that resistance would occur independently of the presence of contaminants. Subsequently, the colonies were divided into a flask or small aliquots in equal proportions and grown independently. The experiment demonstrated that when groups of bacteria are cultured separately, a variable number of specimens resistant to contamination is observed. This finding implied that the mutation occurred before exposure to mud, thereby marking a historic setback for the Lamarckian approach (though see also Holmes et al., 2017).

The revelation of DNA, detailed in Chapter 6 of this volume, provided additional backing for Darwinian theories. It seemed to establish a dominance of genetics over the environment and behavior, affirming the belief that the origin of variability in the phenotype and throughout evolution could be attributed to random mutations of a genetic nature. In 1990, the initiation of the Genome Project (Venter et al., 2001) aimed to attain indisputable knowledge regarding the role played by genes and mutations in causally influencing the onset of human diseases. However, the project's failure, largely tied to impractical reductionist expectations, unequivocally indicated that our complexity, in contrast to other living beings, does not stem solely from the quantity of genes constituting our DNA. Despite articles in *Nature* and *JAMA* prior to the Genome Project's launch suggesting the plausibility of single-gene causation for disorders like schizophrenia (Sherrington et al., 1988) or alcoholism (Blum et al., 1990), this hypothesis resoundingly proved to be untrue. Diseases within a non-linear dynamic system exhibit intricate causes that frequently interact and give rise to potentially adverse outcomes through partially unpredictable pathways. While advancements in human genome sequencing have contributed to enhancing treatments for severe human diseases, including rare ones (Dunbar et al., 2018), the high expectations invested in the Genome Project have largely been disappointing.

Looking at another landscape

During the 1930s and 1940s, a predominantly reductionist perspective in neurobiology was taking shape. However, Conrad Waddington, a British developmental biologist, introduced a groundbreaking theory that would reshape our understanding of the interplay between environment and genetics. While conducting studies on *Drosophila melanogaster*, Waddington (1942) introduced the concept of canalization. This process aimed to explain how traits acquired by one generation through exposure to environmental factors, such as a thermal shock, seemed to be inherited by subsequent generations. Drawing on a concept with roots in Lamarckian ideas, Waddington suggested that canalization allowed potential or latent variations to manifest at the phenotypic level through the inheritance of traits acquired in the environment. In 2015, Noble argued that although Waddington never identified as a Lamarckian, his work provided a mechanism that could be tested against Lamarck's theories. The uniqueness of Waddington's contribution lies in the concept of potential variations, suggesting that not all traits would be inherited across generations. Instead, a set of potentialities, influenced by the environment, could emerge as current phenotypic dimensions (Waddington, 1961).

Due to his significant contributions, Conrad Waddington is widely acknowledged as the pioneer of epigenetics, a discipline focused on examining the mechanisms that, in the interplay between genetics and the environment, give rise to observable phenotypes. In 1957, Waddington introduced the concept of the epigenetic landscape, particularly in the context of embryology. According to this metaphor, a totipotent cell undergoes differentiation over time, traversing a malleable terrain where its movements shape elevations and depressions. In essence, gene-environment interactions dynamically define the epigenetic landscape, generating various possible and probable pathways. Furthering the biological definition of epigenetic mechanisms, Luria proposed that while genetics primarily explores structural modifications affecting cellular genetic material, such as changes in the number or size of nucleic acids, epigenetic modifications entail dynamic alterations in the expression of gene transcription potential.

When environments shape genes

Molecular epigenetics encompasses a diverse array of mechanisms and processes that play a crucial role in regulating transcription. These mechanisms include DNA methylation, histone tail acetylation, phosphorylation, and more. The impact of these changes on DNA transcription is intricate, with a specific epigenetic process potentially resulting in increased or decreased genetic activity based on its location and the surrounding epigenetic context. The interaction among various epigenetic mechanisms gives rise to a complex landscape of epigenetic DNA regulation, referred to as the epigenome. Notably, the epigenome can vary for each cell type in the human body (Bernstein et al., 2007). Understanding many of these epigenetic mechanisms requires recognizing that DNA is not freely present in the cell but is packaged in a structured form known as chromatin, a common configuration in eukaryotic organisms. Chromatin is wound around structural proteins called histones and can be folded in a way that renders it more or less accessible to transcription factors, which enable the synthesis of specific proteins within the cell nucleus. Beyond preventing inadvertent DNA damage and preserving gene content, an essential role of chromatin is to regulate gene expression and cell replication through the action of enzymes known as methyltransferases.

In all vertebrates, the cytosine present in DNA can undergo modification by binding to a methyl group at the carbon element's fifth position. This modification, known as DNA methylation, typically occurs in dinucleotides containing cytosine and guanine, termed CpG sites. The "p" designates the phosphate linking the two nitrogenous bases. Although this mechanism was initially described in a bacterium in the 1920s (Johnson & Coghill, 1925), it was only in the 1980s that it became explicitly associated with gene transcription control processes (Bestor & Ingram, 1983). DNA methylation is an almost ubiquitous epigenetic mechanism in mammals, as up to 80% of cytosines at CpG sites can undergo methylation (Smith & Meissner, 2013). Within the DNA, regions with a high density of CpG sites, commonly known as CpG islands, can be identified. These dense areas often occur near a gene's transcription promotion zone and are typically unmethylated.

The data suggests that "acquired" methylation mechanisms may play a role in regulating chromatin access to transcription, influencing the likelihood of CpG sites within a promoter zone either facilitating or inhibiting the initiation of protein synthesis. In essence, DNA methylation appears to primarily function in inhibiting or silencing gene transcription. However, it's noteworthy that some CpG islands may also be located in intronic regions where methylation could, conversely, stimulate DNA transcription by activating other epigenetic mechanisms (Suzuki & Bird, 2008). Enzymes such as methyltransferases facilitate methylation, and these enzymes can be primarily categorized into two types: de novo methyltransferases, which methylate cytosine nucleotides on both DNA chains, and maintenance methyltransferases, which methylate only one side of the double helix. Maintenance methyltransferases generally act in the opposite position to an already methylated CpG site, intervening in cell replication to maintain the same methylation pattern across different generations of cells.

Modification of histone tails represents a secondary mechanism involved in the epigenetic control of DNA. Histones, being structural proteins with a positive charge, facilitate their interaction with the negatively charged DNA, which is primarily composed of phosphates. Each histone features four elongated tails extending from the chromatin's core to the nuclear environment. This configuration renders the chromatin structure highly susceptible to modifications induced by interactions with numerous proteins (Bannister & Kouzarides, 2011; Yun, 2011). Enzymatic activities play a crucial role in modifying histone tails, with processes such as methylation, acetylation, phosphorylation, and ubiquitination being notable contributors (Kimura, 2013). Acetylation entails the transfer of an acetyl group to lysine residues on the histone tail, promoting gene transcription. This process acts synergistically with demethylation, which is contrary to methylation and reduces gene expression (Szyf et al., 2008). Conversely, deacetylation involves the removal of the acetyl group, leading to a decrease in gene expression, often in conjunction with DNA methylation. Histone tails can also bind to methyl groups, significantly enhancing gene silencing (Ng et al., 2009). Phosphorylation, similar to acetylation, supports gene transcription by attaching a phosphate group to serine and tyrosine residues on histone tails (Tollefsbol, 2011). On the other hand, ubiquitination involves binding the ubiquitin protein to histone terminals, contributing to increased repression of gene expression (Shiio & Eisenman, 2003).

Noncoding RNAs, referring to substantial RNA segments not involved in translating genetic information into proteins, were initially puzzling in terms of their functions. Unlike messenger RNA, noncoding RNA serves as an additional mechanism for regulating gene expression (van der Krol et al., 1990). The exploration of the role of noncoding RNA, especially microRNA, has been a focus in the study of eukaryotic cells (Moazed, 2009). MicroRNA is capable of altering chromatin structure through RNA interference processes, leading to gene silencing (Rinn & Chang, 2012). While the epigenetic regulatory machinery comprises various synergistic mechanisms, DNA methylation stands out as the most extensively studied epigenetic mechanism in many animal models and humans. This prevalence arises from the ease with which DNA methylation can be assessed, both in specific DNA segments identified as targets based on previous studies and at the level of the entire methylome (Giorda, 2020). Consequently, the next section of the chapter will predominantly focus on this epigenetic process. It will illustrate how, originating from experiments conducted on animal models, methylation has become synonymous with epigenetic modifications in a significant portion of the scientific literature in developmental psychobiology (Meaney & Szyf, 2005).

Stories of rodents

The study of epigenetic mechanisms has become particularly intriguing for researchers focused on understanding the development and behavior of humans and animals. This interest stems from the potential that many of these processes might be influenced, at least partially, by environmental factors (Kang et al., 2011; Maze et al., 2014). Current knowledge indicates that adverse experiences during early pre- and postnatal stages can alter the methylation levels of specific genes involved in stress regulation (Mulligan et al., 2012; Perroud et al., 2014; van der Knaap et al., 2014; Tyrka et al., 2012)—a topic discussed more comprehensively in Chapter 11.

However, the foundational studies paving the way for behavioral epigenetics, exploring the consequences of environmental stimuli-induced epigenetic variations on behavioral development, were initially conducted on animal models, particularly laboratory rats and mice (Berretta et al., 2021). As previously mentioned in Chapter 5, rodents have been extensively utilized to investigate the short- and long-term impacts of parental care behavior on the behavioral, cognitive, and neuroendocrine development of offspring. Similar to humans, rodents heavily rely on early care for their growth and increased chances of survival. Rat mothers exhibit immediate responsiveness to their offspring's signals (Rosenblatt et al., 1988). Immediately post delivery, mothers keep the offspring close, facilitating temperature regulation and providing tactile stimulation. While the constellation of maternal behavior in rats is less intricate than in humans, it still offers variability in the frequency, duration, and quality of observable and quantifiable behaviors such as licking-grooming (LG) or arch-backed nursing (ABN) (Berretta et al., 2021). These behaviors, widely present in the rat mother population, show significant individual differences even in lab-raised animal colonies. Studies conducted by Michael Meaney and colleagues on Long-Evans rats, where mothers exhibit varying tendencies toward LG-ABN behaviors, have emphasized how physiological variations in maternal behavior may be associated with distinct evolutionary outcomes for offspring. These investigations underscore the involvement of epigenetic mechanisms in linking changes in maternal behavior to the regulation of offspring stress.

Rats born to mothers exhibiting a high frequency of LG and ABN behaviors, denoted as the high-quality maternal care group (HQC), displayed distinct physiological and behavioral characteristics. This included a reduced elevation of ACTH and corticosterone in response to laboratory stress, increased expression of hippocampal receptors for glucocorticoids, more efficient regulatory feedback action of the HPA axis, and lower CRH expression in the hypothalamus (Liu et al., 1997). At the behavioral level, offspring from HQCHQC mothers appeared less fearful in the presence of new stimuli (Caldji et al., 1998) and demonstrated enhanced spatial memory in Morris' water maze, a commonly used task for assessing learning in laboratory rats (Liu et al., 2000). To discern whether these differences were attributed to genetic factors or postnatal exposures, researchers employed a cross-fostering procedure. This involved removing the young from their original litter and having them raised by an adoptive mother. The objective was to investigate whether the observed distinctions between offspring from HQC mothers and those from mothers displaying low-quality care (LQC) were a result of hereditary factors or environmental influences. Through these investigations, it was established that the behavioral and physiological variances observed in the offspring of HQC and LQC mothers were associated with postnatal manipulations (Champagne et al., 2003; Weaver et al., 2004). Specifically, when offspring born to LQC mothers were raised by HQC "adoptive" mothers, they exhibited a behavioral phenotype in adulthood that was indistinguishable from their counterparts born and continuously cared for by HQC "biological" mothers.

In 2004, Weaver and colleagues significantly reshaped our understanding of parenting and the interplay between genetics and the environment with their article titled "Epigenetic programming through maternal behavior." This pivotal work expanded the scope of behavioral epigenetics beyond laboratory animal models, paving the way for studies conducted in humans and children (Hyman, 2009). The authors provided a comprehensive description of the mechanisms involved in behavioral epigenetic regulation. Initially, they identified a specific gene, NR3C1, responsible for coding hippocampal glucocorticoid receptors (GR). In offspring born and cared for by low-quality care (LQC) mothers, this gene exhibited high methylation, while offspring from high-quality care (HQC) mothers showed minimal methylation. Through a limited cross-fostering assignment of some pups from LQC mothers to HQC mothers, the researchers observed that methylation of the NR3C1 gene's promotion region (exon 1F) depended on the quality of care received after the cross-fostering. Essentially, the epigenetic regulation of the gene governing GR expression appeared to be under at least partial control of postnatal maternal behavior. The authors further examined methylation levels of the same gene at various ages of offspring. The significant and consistent differences between the HQC and LQC groups endured throughout the lifespan into adulthood, although they were not present during the prenatal and neonatal periods.

In addition, HQC offspring exhibited greater acetylation of histone H3-H9 tails and up to three times more binding between the neural growth factor NGFI-A and the 1F-promoting region of the NR3C1 gene in the hippocampus. These findings indicated increased transcription of the gene in this brain region, emphasizing the role of postnatal maternal behavior in the epigenetic regulation of gene expression.

Additional evidence indicates that the prenatal maternal environment, particularly the diet during pregnancy, can exert a significant influence on the baby's development through epigenetic modifications. In the late nineties, Wolff and colleagues (1998) conducted a study on Agouti mice. These mice, characterized by a yellow coat, tend to be obese and are prone to diseases like diabetes and cancer. In contrast, pseudo-agouti mice display a brown coat, a lower risk of developing similar pathologies, and notably, a higher level of methylation in the promoter of the agouti gene, responsible for coat coloration in this species. Wolff's group demonstrated that environmental factors related to maternal diet modulated the epigenetic regulation of this gene. Specifically, when the mother was fed a diet rich in methyl group donor agents, such as folic acid, vitamin B12, or choline, the offspring was more likely to exhibit a pseudo-agouti phenotype.

Collectively, these studies have transformed our understanding of the environmental impact on genetic regulation, especially in the context of DNA function, encompassing transcription and protein synthesis, crucial processes that shape behavior and development. While other factors may influence the efficiency of cell replication activity (as discussed in the next chapter regarding telomere regulation), these factors also seem susceptible to environmental influences. However, functional methylation modifications appear to maintain stability over time, preserving the phenotype acquired from the environment. This stability contributes to the long-term orchestration of adaptive development in the child.

Hallelujah

I often use a metaphor involving music to illustrate how epigenetic processes, especially methylation, function. Consider DNA as the musical score of a well-known song, say, "Hallelujah" by Leonard Cohen. The sequence of notes in the score is akin to the sequence of DNA dinucleotides. When you play the song on the piano, your hands act as a motor, transposing the two DNA chains and connecting the notes of the melody with the accompaniment played by the left hand. Now, the notes on the score are fixed, black on white; you can see them, identify them, and name them. However, once you start playing, the music produced becomes your unique interpretation of that score. In other words, you can follow the given instructions—the musical notations indicating tempo changes, dynamics (piano, pianissimo, forte, fortissimo)—and at the same time, inject your own subjective interpretation regarding pauses, intensity, and rhythm. This parallels what happens with Leonard Cohen's "Hallelujah," which has been reinterpreted by various artists such as Jeff Buckley, Elisa, Damien Rice, Rufus Wainwright, and many others. Listening to different performers' versions of this song reveals how, drawing on their individual sensitivity, each artist has provided a distinct rendition of "Hallelujah."

Taking the metaphor further, we can imagine that the sensitivity of each artist reflects their genetic predisposition to varying degrees of sensitivity to certain intensities or rhythms of stimulation. However, equally significant is the influence of life experiences and the environment in which they grew up and developed, both as individuals and musicians.

Similarly, it's conceivable that as a result of the inevitable life experiences we undergo and the caregiving and relational environments we navigate and choose, each of us can metaphorically contribute to the musical score of our DNA. Through epigenetic processes, we may insert new notations that instruct our genetics on how to behave under specific circumstances—facilitating the expression of certain gene portions and inhibiting others. In this perspective, DNA ceases to be merely a rigid set of instructions to be strictly followed, capable of determining an individual's fate. Instead, it transforms into a digital narrative, adaptable and amendable with the memories of our journeys and relationships. From this viewpoint, our DNA becomes a dynamic story, reflecting the signs and scars of our past experiences, influencing the regulation of our behavior in the present. It represents a deliberate, albeit non-conscious, intentional process of epigenome regulation. This intentional shaping of our genetic expression occurs beyond the realm of awareness or explicit decision-making, demonstrating the intricate interplay between our experiences, choices, and the regulation of our genetic behavior.

Just as certain passages in a musical score lend themselves to subjective interpretations, specific portions of DNA appear more intriguing than others in the realm of epigenetic regulation. Two genes, in particular, have undergone extensive study for their impact on the cognitive, behavioral, and socio-affective development of children: the gene regulating the expression of hippocampal GRs (NR3C1) and the gene coding for the serotonin transporter (SLC6A4). The roles of serotonin and glucocorticoids (cortisol in humans) have been explored in previous chapters, emphasizing their significance in regulating behavioral states during challenging and stressful situations. Their involvement in reactivity, stress recovery, and mechanisms related to learning and memory makes them particularly relevant.

Building on animal model studies, researchers have investigated how adverse environmental exposures may influence the regulation of the NR3C1 gene in humans (Palma-Gudiel et al., 2015). Early adverse experiences appear capable of modifying the epigenetic regulation of this gene, notably by elevating methylation levels, leading to progressive gene silencing (Conradt et al., 2019; Oberlander et al., 2008). A recent meta-analysis (Berretta et al., 2021) indicates that early adverse experiences, occurring in the prenatal or postnatal period within the first thousand days of conception, are consistently associated with increased methylation of the gene encoding cortisol receptors (Kantake et al., 2014; Oberlander et al., 2008). Moreover, specific maternal behaviors, such as breastfeeding (Lester et al., 2018), social touch (Murgatroyd et al., 2015), and maternal sensitivity (Conradt et al., 2016), are linked to a decrease in the level of methylation of the NR3C1 gene, aligning with findings from animal models using the cross-fostering paradigm (Weaver et al., 2004). The long-term effects of observed epigenetic variations on a child's behavioral development are less conclusive. Some studies suggest an association between increased methylation levels and lower temperamental regulation skills (Appleton et al., 2015; Paquette et al., 2015), while others report the opposite effect (Bromer et al., 2012; Folger et al., 2019; Stroud et al., 2016), indicating the involvement of additional mechanisms in determining a child's developmental trajectories.

There is substantial evidence supporting the role of the serotonergic system in modulating stress reactivity (Lesch, 2011), and recent studies propose that the SLC6A4 gene may be particularly susceptible to environment-related epigenetic regulation (Booij et al., 2013; Provenzi et al., 2016b). The gene encoding the serotonin transporter is known to be genetically regulated through a polymorphism involving two allelic variants: a short variant (S) producing less transporter and a long variant (L) producing more transporter (Lesch, 2011). The short variant has been associated with alterations in limbic system functioning (Hariri et al., 2002) and increased susceptibility to stress in humans and young children (Montirosso et al., 2015; Pauli-Pott et al., 2009; Pluess et al., 2011). However, genetic research on the SLC6A4 polymorphism indicates that variations in phenotype associated with 5-HTTLPR genetic variants are more complex than explained by genotype alone. Studies in macaques have shown that early adverse experiences may be linked to reduced SLC6A4 gene expression and uninhibited social behaviors in stressful laboratory sessions (Kinnally et al., 2008). A recent meta-analysis of 19 studies (Provenzi et al., 2016) documented that methylation of the SLC6A4 gene in humans can reliably serve as an index of early (prenatal and postnatal) and late (adolescence) exposure to stress.

Behavioral epigenetics

The studies conducted by Meaney and colleagues have given rise to a novel discipline known as behavioral epigenetics. This field seeks to apply epigenetic principles to explore the physiological, genetic, behavioral, and environmental mechanisms involved in the psychobiological development of both humans and nonhuman animals (Lester et al., 2011). Specifically, the focus is on understanding how certain behavioral outcomes can be associated with epigenetic precursors regulated concurrently or as a consequence of specific experimental or quasi-experimental manipulations in the early living environment. Behavioral epigenetics investigates how genes, or more precisely, the mechanisms regulating their expression, can be influenced by environmental stimuli, contributing to the programming of health or disease trajectories throughout life (Lester et al., 2016). This discipline holds promise for significant advancements in the study of substance dependence disorders (Maze & Russo, 2010), memory and learning processes in typical development (Day & Sweatt, 2011), neurodevelopment, parental behavior (Meaney & Szyf, 2005), psychiatric pathologies (Tsankova et al., 2007), and stress regulation in general (Provenzi et al., 2016).

By providing an observable and quantifiable mechanism for the transmission between the environment and the transcriptome (Champagne, 2008), behavioral epigenetics appears to offer new insights into individual variability in resilience or the development of stress-related diseases (Yehuda & Bierer, 2009). Over the past decade, the number of studies in behavioral epigenetics has seen exponential growth, fostering a specific interest in developmental psychobiology regarding the impact of early adverse experiences on child development. In essence, numerous studies have been initiated to comprehend how early stressful and adverse events—encompassing neglect, abuse, maltreatment, as well as natural disasters and chronic stress exposure—can leave a lasting imprint on the development of children with typical development and developmental risk (Provenzi & Montirosso, 2020). While detailed examples of how epigenetic modifications may play a pivotal role in the psychobiological, behavioral, social, and cognitive development of children will be presented in the chapters of the third section of this book, the following pieces of evidence offer an initial understanding of the profound impact of the initial animal model studies on our field of investigation.

McGowan and colleagues (2009) conducted the initial study that adopted Meaney and colleagues' conceptual model, revealing elevated methylation levels and reduced expression of the NR3C1 gene in individuals who died by suicide with a history of abuse compared to suicides without a history of abuse and controls who died from natural causes. A study examining global methylation in the hippocampus identified more than 300 regions displaying differential methylation between individuals reporting a history of childhood abuse and control subjects. Nearly 70% of these regions exhibited hyper-methylation. In war veterans, those who developed posttraumatic stress disorder (PTSD) displayed lower methylation of the NR3C1 gene in the blood, with these levels predicting a favorable response to psychotherapy (Yehuda et al., 2013). Studies utilizing non-localized epigenetic investigations, known as epigenome-wide approaches, demonstrated that individuals with PTSD exhibited significantly different methylation levels in gene regions related to the response to inflammation (Smith et al., 2011).

Notably, extensive research has been conducted on the effects of the Dutch Famine, a human-induced famine occurring in the 1940s toward the end of World War II. The famine, resulting from a Nazi embargo on food transport in West Holland after the Allies' attempt to liberate the southern part of the country, led to food rations in Amsterdam falling below 800 kcal (Stein et al., 1975). Studying the long-term, cross-generational effects of the *Hongerwinter* allowed researchers to discover that, six decades later, the grandchildren of those exposed to famine exhibited less methylation of genes involved in fetal growth and maternal imprinting (Heijmans et al., 2008). Women who developed posttraumatic stress symptoms due to the September 2001 Twin Towers attack gave birth to children showing reduced expression of the FKBP5 gene, implicated in regulating stress-related inflammatory processes (Yehuda et al., 2009).

While most behavioral epigenetics studies in humans often involve retrospective experimental designs, aiming to understand the impact of adverse life experiences on epigenetic regulation and adaptive behavior, some researchers initially sought to replicate the early results of Meaney and Szyf (Oberlander et al., 2008). Due to limited possibilities for manipulating the caregiving environment in the human model, controlled prospective studies are restricted to specific conditions, such as preterm birth or quasi-experimental situations like natural disasters or large-scale emergencies. Consequently, the exploration of epigenetic correlates of protective or therapeutic experiences is quite limited in humans, unlike the animal model where direct manipulation through the cross-fostering paradigm is feasible.

Nevertheless, a subset of studies has explored how physiological variations or risk conditions in maternal behavior may be linked to epigenetic regulation profiles influencing the child's development in the first years of life. A systematic literature review identified eleven articles comprising a total of 1399 subjects. Six studies demonstrated a significant association between changes in the quality of parental behavior and methylation profiles observed in genes such as NR3C1 and SLC6A4, as well as other genes involved in cognitive and social development like BDNF and the gene encoding the oxytocin receptor (OXTR). In addition, in some studies, maternal behavior appeared to play a significant buffering role against potential adverse effects of early stressful experiences (Conradt et al., 2016; Murgatroyd et al., 2015; Provenzi, Fumagalli et al., 2016).

The seductive charm of a promise

On the internet, you may come across a comic strip that humorously suggests, "If there is something you do not know, answer that it is epigenetic." While intended as irony, this statement carries a kernel of truth. The field of behavioral epigenetics captivates researchers and clinicians, offering a seemingly straightforward answer to questions surrounding the origins and mechanisms of psychopathology: the environment regulates genetics. However, as discussed earlier, there are numerous constraints and unanswered challenges in applying epigenetics to the study of psychobiology and developmental psychopathology. One might view epigenetics as a discipline still in its early stages, holding the promise of becoming a valuable tool but falling short of providing a definitive explanation for the intricate interplay between nature and culture (Austerberry & Fearon, 2020). Acknowledging these limitations and addressing the unresolved challenges is crucial for advancing and strengthening the field of behavioral epigenetics in the study of human development.

In 2010, Miller published an article in Science with the emblematic title "The Seductive Allure of Behavioral Epigenetics." While he praised the potential implications of Meaney's studies for public health and prevention, he also issued a cautionary note against the temptation of viewing DNA methylation as the new Holy Grail capable of conclusively explaining, and perhaps oversimplifying, all events related to both typical and pathological human behavior. Likewise, Richardson and colleagues (2014) warned about the potential risk that the growing confidence in epigenetic mechanisms as explanatory factors might lead to a renewed wave of blame directed at mothers. This blame could arise from the perception that mothers are responsible for the epigenetic dysregulation of their children, resembling a psychobiological revival of the concept of a "refrigerator mother" or "schizophrenogenic mother." Furthermore, in a 2015 article titled "Epigenethics," published in *JAMA Pediatrics*, Provenzi and Montirosso (2015) underscored this aspect by coining the term, a blend of epigenetics and ethics. The term was coined to emphasize the necessity of utilizing data obtained from this emerging line of studies in a rigorous, scientific, clinical, and ethical manner. This highlighted the importance of avoiding unwarranted conclusions or stigmatization, ensuring that ethical considerations accompany the advancements in our understanding of epigenetics.

The limitations of studies in behavioral epigenetics in humans are diverse. First, the retrospective nature of many such studies hinders their ability to provide causal interpretations of highlighted significant associations (Provenzi et al., 2017a). Furthermore, some studies concentrate on examining individual genes, while others prefer employing epigenome-wide approaches, analyzing a broad array of potential methylation sites without a specific target. These methods pose distinct challenges. On one hand, the single-gene approach may struggle to identify robust and clear associations explaining a behavioral trait or developmental outcome based on a few CpG sites. On the other hand, the epigenome-wide approach opens the door to discovering new associations and optimizing analysis costs, but there remains a need to offer a biologically and clinically meaningful interpretation of statistically significant associations. Both methods can address different questions and are applicable in contexts with varying resources. Moreover, in humans, epigenetic data is typically derived from peripheral tissues, frequently blood or saliva. In contrast, in animals, methylation quantification is assessed in central tissues—directly in the hippocampus or brain regions of interest—after sacrificing the animal. At present, conflicting data exist regarding the consistency of methylation level measurements in different tissues within the same subject, with variations potentially dependent on the investigated gene (Provenzi, Brambilla et al., 2018).

Simultaneously, the significance and potential of this field of study are evident. Despite being in its early stages, the application of epigenetic research and its principles to comprehend the mechanisms governing human psychobiological development holds great promise. Prospective longitudinal studies aimed at exploring the epigenetic connections of various early preventive or therapeutic interventions could not only unveil the pathways by which a treatment might prove effective or not but also pinpoint individuals who stand to benefit from different interventions based on the regulatory status of their epigenome. As the demand for early interventions continues to rise in a socio-health context marked by limited resources, the ability to identify the most effective and efficient responses for different patients becomes a priority, and behavioral epigenetics can play a decisive role in contributing to this objective.

8 Other frontiers

In the earlier sections of this volume, we have delved into pivotal mechanisms that have contributed to the evolution of developmental psychobiology as a research domain over the years. Each mechanism or biomarker offers only a glimpse into the intricate regulation—both in terms of order and disorder—of the complex non-linear dynamic system that constitutes the human being in its developmental journey and interaction with the environment. As knowledge progresses and innovative techniques for probing physiological and biological processes related to stress regulation and socio-emotional and socio-cognitive behavior come to the forefront, fresh measures enable us to perceive the complexity of human beings from diverse perspectives. The objective of this chapter is twofold. First, it aims to kindle the reader's curiosity and prompt exploration of new psychobiological processes that can contribute to understanding how everyday experiences shape our identities. Second, it advocates for the acknowledgment of psychobiological complexity, urging caution against reductionist approaches that may emerge from an excessive reliance on a limited set of processes with the simplistic belief that they alone can adequately explain the non-linear dynamics of human development.

Stress that can inflame

While the hypothalamic-pituitary-adrenal (HPA) axis system and the sympathetic nervous system are commonly recognized as the primary mediators of the body's stress response, both glucocorticoids and catecholamines exert indirect and widespread effects on various target systems. One of these systems is the immune system, and there is a recent hypothesis suggesting that pro- or anti-inflammatory processes play a significant role in our ability to regulate stress, especially chronic stress (Rohleder, 2014). The immune system demonstrates particular sensitivity to psychosocial stress and can be activated by neuroendocrine mediators involved in stress regulation and immediate reactivity, such as catecholamines and glucocorticoids. In addition, immune cells mediate inflammatory processes and communicate their activity to the central nervous system.

Cytokines serve as pivotal components in the inflammatory immune response, typically released in reaction to infection and tissue damage. Various cell types release these signaling molecules, indicating the need for activating an immune

DOI: 10.4324/9781003479314-10

response at the intercellular level. Much like glucocorticoids, cytokines interact with specific receptors on the surface of target cells, enabling communication over long distances throughout the body. Broadly, cytokines can be categorized into pro-inflammatory and anti-inflammatory types. Pro-inflammatory cytokines, including interleukins (IL) such as IL-1 and IL-6, along with tumor necrosis factor alpha (TNF-α), are produced by cells situated in regions of infection or damage. Conversely, anti-inflammatory cytokines like IL-4, IL-5, IL-10, and IL-13 participate in the anti-inflammatory process, mitigating the immune response, inhibiting immune activity, and suppressing the synthesis of other cytokines. Certain pro-inflammatory cytokines can directly impact the brain, stimulating the activity of the HPA axis and triggering a neuroendocrine stress response. Also, glucocorticoids, the primary effectors of the HPA axis, play a role in initiating immunosuppressive and anti-inflammatory actions. They diminish the expression of pro-inflammatory cytokines such as TNF-α and IL-6, while promoting the expression of anti-inflammatory cytokines such as TNF-β and IL-10.

Exposure to acute or brief stress triggers the initiation of an inflammatory response characterized by an increase in pro-inflammatory cytokines. Cytokines, a diverse class of highly heterogeneous proteins, regulate the immune response through intercellular signaling (Robles et al., 2005). Pro-inflammatory cytokines, in particular, play a role in mobilizing an immune response in reaction to psychosocial stress. However, prolonged and excessive production of pro-inflammatory agents can have detrimental effects on the body, elevating the risk of conditions such as depression, posttraumatic stress disorder (PTSD), and various psychosomatic disorders (Robles et al., 2005).

For example, at low concentrations, IL-6, IL-1beta, and TNF-α promote neural differentiation and nerve cell proliferation. Conversely, at high concentrations, they hinder neurogenesis and jeopardize cell survival (Arauko & Cotman, 1995; Bernardino et al., 2008; Zunszain et al., 2012). Individuals diagnosed with major depressive disorder often exhibit elevated levels of cytokines, particularly interleukins such as IL-6 and IL-1β, TNF-α, and C-reactive protein (CRP) (Dowlati et al., 2010; Liu, Ho, & Mak, 2012). In 10-year-old children, heightened levels of behavioral problems are associated with elevated CRP and IL-6 levels (Slopen et al., 2013). Despite cytokines not directly crossing the blood-brain barrier, there is a possibility that they exert their effects through indirect pathways (Dantzer, 2009).

CRP, produced in the liver and classified as a pro-inflammatory cytokine, plays a pivotal role in acute responses to infections and stressful events. High concentrations of CRP have been linked to psychological disorders such as depression and PTSD (Gill et al., 2009; Valkanova et al., 2013). Elevated CRP levels have been observed in individuals exposed to chronic stress conditions, including those who experienced terrorist attacks in Israel (Canetti et al., 2014) or traumatic events like survivors of the World Trade Center attack (Rosen et al., 2017). In both studies, higher CRP levels were associated with more severe posttraumatic conditions. Longitudinal studies have shown that early experiences of maltreatment predict significantly increased plasma CRP levels in adulthood (Danese et al., 2007). These findings regarding CRP have been replicated by other researchers (Appleton et al., 2012; Pollitt et al., 2007), and significant effects have also been demonstrated for IL-6, another cytokine (Bertone-Johnson et al., 2012).

Carpenter and colleagues (2010) observed that the immune response, specifically IL-6, to psychosocial stress in a laboratory setting was more pronounced in adults with a history of early maltreatment. In addition, they found an effect on baseline levels of TNF-α, IL-6, and IL-1β a few years later (Hartwell et al., 2013). IL-6 has been identified in high amounts in adults with depressive pathology and exhibits a significant response to antidepressant drug treatments (Kenis and Maes, 2002; Glaser et al., 2003). In a more recent study by Tyrka and colleagues (2015), children aged between three and five years, exposed to maltreatment, were assessed for salivary levels of CRP and IL-1β. Although no associations were found between the number of adverse events and CRP concentrations, a significant effect emerged for IL-1β, a cytokine previously linked to stress exposure in animal models (Caso et al., 2007; Nguyen et al., 1998) and in response to conditions of social or cognitive stress in humans (Brydon et al., 2005; Mastrolonardo et al., 2007).

In children, psychosocial stress, particularly related to the quality of the relationship with caregivers, seems to impact the activity of factors involved in inflammatory processes. In a sizable sample of 4-month-old macaques separated from their maternal figures for 2 hours, higher levels of pro-inflammatory cytokines were observed in young mothers assessed as having lower levels of maternal sensitivity (Kinnally et al., 2019). A research group led by Pariante at King's College London investigated CRP levels in 25-year-old adults who had been followed longitudinally from birth (Plant et al., 2016). In this cohort, exposure to prenatal depression significantly predicted CRP levels at age 25, independent of postnatal stressful events such as maltreatment and depression in adulthood. More recently, 12-month-old babies with a secure attachment style exhibited lower salivary CRP levels compared to age-matched infants with insecure attachment 6 months later (Nelson et al., 2020). In a second study, the same authors found that the social support perceived by mothers was associated with the increase in CRP observed from 12–18 months: mothers who reported higher levels of social support had children who showed lower levels of pro-inflammatory cytokines at 18 months (Nelson et al., 2020).

A second brain

Today, the recognition that microbes residing in the human gut can significantly influence metabolism and potentially contribute to the discovery of new therapies might be self-evident. However, this insight has only become achievable through recent advancements in genetics and epigenetics. At present, we understand that all multicellular organisms play host to microbial life, and this enduring relationship between microorganisms and their hosts has persisted over millions of years of evolution (Moeller et al., 2016; Nishida & Ochman, 2018). The microbiome, an expansive and intricate realm of microorganisms, deeply shapes the health of the organism and engages with various cell types within our bodies. In mammals, a pivotal moment of host colonization takes place during delivery, transferring maternal microbes to the newborn's gut (Ferretti et al., 2018; Sprockett et al., 2018). Following this, the microbiome consistently undergoes rearrangements in response to environmental cues, health conditions, and developmental processes (Marchesi & Ravel, 2015). Childhood emerges as a particularly vulnerable period for microbiome establishment, transitioning from a comparatively sterile environment to a complex ecosystem housing over 3×10^{13} species of microorganisms (Sender et al., 2016).

Various methodologies are employed to investigate the intricate interaction between the microbiome and the host organism. These approaches encompass exogenous manipulations, involving drugs, antibiotics, probiotics, or prebiotics. In addition, studies utilize germ-free animals raised in sterile environments, and the transfer of microbial cultures is facilitated through fecal samples in co-housing scenarios. Insights garnered from research utilizing these approaches have unveiled the substantial role of microbes in regulating diverse aspects of the host organism's physiology, including metabolism, adiposity, energy balance, appetite, and nutrition. Moreover, the microbiome exerts influence over the maturation and activity of the immune system. More recent speculation suggests a potential impact of the microbiome on brain development (Sharon et al., 2016). However, the current comprehension of the mechanisms through which the microbiome affects the central nervous system largely stems from research conducted on animal models. Credible candidates for understanding this association include metabolite production, interactions with the immune system, and regulation via the vagus nerve (Sherwin et al., 2019). Studies involving germ-free rats have contributed valuable insights into the influence of the microbiome on brain development and activity. These animals exhibit distinctive characteristics such as increased myelination in the frontal cortex, immature microglial development, aberrant neurogenesis, altered brain volumes in regions like the neocortex and amygdala, and heightened permeability of the blood-brain barrier (Hoban et al., 2016). Notably, these rats also display an elevated risk of internalizing behavioral disorders, including isolation and a diminished capacity to develop adaptive responses to fear and anxiety (Chu et al., 2019).

Moreover, alterations in the gut microbiome have been linked to cognitive development (Carlson et al., 2018), functional connectivity in the central nervous system (Gao et al., 2019), stress response, and amygdala volume (Carlson et al., 2018). Early modifications in the microbiome are particularly pertinent, as they may contribute to dysregulation in stress regulation by the HPA neuroendocrine system. Research on germ-free rats has demonstrated hyper-reactivity of the HPA axis in response to physical stimulation and social stress conditions comparable to those encountered by humans (Crumeyrolle-Arias et al., 2014; Sudo et al., 2004). In humans, specific prebiotics and probiotics, such as lactobacillus, have been associated with reduced salivary cortisol levels and other stress indicators (Messaoudi et al., 2011). The stress response during pregnancy has been positively correlated with the abundance of Rikenellaceae and dialister, and negatively with the percentage of bacteroides (Hantsoo et al., 2019). However, the evidence supporting these effects is not yet robust and requires further validation.

A body of evidence suggests the involvement of the microbiome in social and psychological processes, encompassing the regulation of emotions, learning, and memory (Montiel-Castro et al., 2013; Sarkar et al., 2020). Multiple potential mechanisms exist through which the microbiome could influence socio-emotional and socio-cognitive functioning in mammals. First, the microbiome has been found to impact the development of brain regions involved in emotion regulation and memory, such as the amygdala, hippocampus, and prefrontal cortex (Sudo et al., 2004). Also, the microbiome can modify the bioavailability of various molecules that, in turn, influence social behavior, including glucocorticoids, neuropeptides, and catecholamines (Poutahidis et al., 2013; Wikoff et al., 2009).

Research conducted on germ-free mice and rats using the three-chamber paradigm, which assesses preferences for previously unmet conspecifics, reveals deficits in social behavior. These animals do not exhibit a preference for other conspecifics when encountering unfamiliar or familiar objects (Crumeyrolle-Arias et al., 2014; Stilling et al., 2018). In addition, the microbiome has been found to influence gene expression and epigenetic processes relevant to affiliative behavior (Sarkar et al., 2020). Correlational studies in animal models indicate an association between social stress and alterations in the microbiome. Stress induced by behaviors such as social aggression and subordination to a dominant conspecific leads to changes in the gut microbiome and immune function in mice (Bharwani et al., 2016). Isolation and maternal separation have also been linked to changes in the abundance of specific bacterial taxa (Dunphy-Doherty et al., 2018). Resilient mice to stress exhibit a higher prevalence of Bifidobacteria compared to susceptible individuals (Yang et al., 2017). Neuroimaging studies in germ-free mice reveal a greater amygdala volume, reduced prefrontal cortex thickness, and diminished hippocampal expression of the neurotrophic factor BDNF (Clarke et al., 2013; Hoban et al., 2016). In humans, some probiotics – such as the lactobacillo – might diminish the levels of salivary cortisol as well as other markers of stress (Messaoudi et al., 2011)

Studies on children indicate that breastfeeding is associated with the transmission of prebiotics to the baby's gut (Charbonneau et al., 2016). Early exposure to antibiotics in childhood may permanently alter the gut microbiome and endocrine physiology in rodents, and such children may exhibit a higher incidence of depressive symptoms (Cho et al., 2012; Slykerman et al., 2015). The infant microbiome can also be influenced by early prenatal stressful events, with higher levels of maternal stress during pregnancy associated with changes in the baby's microbial composition (Zijlmans et al., 2015). Moreover, greater variability in the microbiome bacterial flora in children has been associated with different measures of behavioral and emotional development (Aatsinki et al., 2019; Carlson et al., 2018; Christian et al., 2015; Loughman et al., 2020). However, there are diverse and partially conflicting associations, and future research needs to focus on specific microbes or the genes they express.

Although the mechanisms by which the microbiome influences psychological and social development are not fully understood, it is reasonable to anticipate a pervasive effect on various physiological and biological regulatory processes, considering that animal life evolved in the presence of microorganisms (McFall-Ngai et al., 2013). Human studies suggest that the microbiome may play a role in social-emotional regulation, but further research is needed to validate these findings and understand the environmental and neurobiological mediators at play during early childhood and later development.

Powerful organelles

Mitochondria are intracellular organelles thought to have originated from an endosymbiotic relationship. Two main evolutionary hypotheses exist regarding the origin of mitochondria. The first proposes that around two billion years ago, a bacterium resembling mitochondria was engulfed by an anaerobic cell with a nucleus, leading to the emergence of the future eukaryotic cell. This hypothesis suggests that symbiosis initially provided the eukaryotic cell with the ability to detoxify oxygen, considering the original bacterium as an obligate aerobic unicellular organism. However, this view raises questions, as the toxic forms of oxygen for anaerobic organisms are actually products of mitochondria in eukaryotic cells, and it assumes the endosymbiont cell already had a nucleus. A second, more plausible hypothesis suggests that the host cell incorporating mitochondria was a prokaryotic cell or a true archaeobacterium. According to this model, the ancestral mitochondria were optionally anaerobic and could exist with or without oxygen availability. Symbiosis in this scenario would have been advantageous due to the production of hydrogen by the endosymbiont as an energy source and electrons for the embedded mitochondria. This hypothesis may have given rise to the eukaryotic cell as a genetic chimera.

Mitochondria play a crucial role in sustaining human life by generating energy and intracellular signals. With the exception of oxygen-carrying red blood cells, all cells in the human body house hundreds or thousands of mitochondria. These organelles undergo continuous fission or fusion processes, making their quantification challenging due to their rapid and dynamic reactions. A distinctive feature of mitochondria is their possession of their own genome, known as mitochondrial DNA, which codes for genes essential in the production and transport of energy through the electron transport chain. This chain, operating in the mitochondrial membrane and consuming oxygen, generates an action potential. Much like batteries storing electrical charges for powering devices, organisms "breathe" to introduce energy and charge their mitochondria. In mitochondria, this charge is harnessed to produce additional energy in the form of adenosine triphosphate (ATP), supporting various systems, including neural, cardiac, muscular, digestive, and those involved in the stress response.

Mitochondria exhibit the capacity to sense stress-related mediators or signals within the body, including glucocorticoids, estrogen, angiotensin, and cannabinoids. Similar to synapses in nerve cells, inter-mitochondrial junctions enable the exchange of information in a distributed network of mitochondria responsive to stress signals. Prolonged exposure to these stress signals can induce morphological and functional alterations in mitochondria, potentially leading to allostatic recalibration processes and the accumulation of damage at the mitochondrial DNA level (Picard et al., 2013). These changes in mitochondrial function can generate signals influencing cellular functioning through epigenetic mechanisms. Various aspects of human genome functioning are under mitochondrial control (Picard & McEwen, 2018), and all steroid hormones, including glucocorticoids, are synthesized through processes regulated by mitochondria (Midzak & Papadopoulos, 2016). Genetic variations in mitochondrial DNA have been shown to alter corticosterone production in response to stress in mice (Gimsa et al., 2018), and individuals with mitochondrial mutations associated with increased exposure to oxidative stress may display hypocortisolemia (Meimaridou et al., 2012). In addition, patients with mitochondrial disorders have exhibited a decrease in vagus nerve activity in response to motor activity (Bates et al., 2013).

The mitochondria appear to play an important role in regulating the inflammatory immune response to stress as well. When the body is exposed to oxidative stress, mitochondria produce proteins – called alarmins – that signal their status to the rest of the body and the central nervous system. This occurs because mitochondria, being of bacterial origin, are recognized as foreign by the immune system (Lu et al., 2014). As a result, cytokine levels increase, and inflammatory processes occur, observed in animal models and humans (Oka et al., 2012).

In addition to alarmins, another mitokine – another pro-inflammatory factor derived from mitochondria – is free mitochondrial DNA, a fraction of mitochondrial epigenome in the liquid fraction of blood, particularly present in inflammatory disorders associated with serious illnesses such as cancer (Boyapati et al., 2017), and in individuals with a history of attempted suicide (Picard et al., 2018). At the same time, mitochondria can also influence telomeric erosion regulation processes. Reactive oxygen species produced by mitochondria can cause telomere shortening in vitro, and cellular senescence seems to be inhibited in the absence of mitochondria. In humans, the number of copies of mitochondrial DNA decreases with age (Mengel-From et al., 2014) and correlates significantly with telomere length (Pieters et al., 2015). Furthermore, the association appears to be stronger in individuals with a history of maltreatment and psychopathology (Ridout, Carpenter, Tyrka, 2016). It is possible that some inflammatory processes – such as IL-6 activity – are involved in mediating the association between mitochondrial activity – for example, respiratory activity – and telomere erosion (Boeck et al., 2016).

In this way, mitochondria also contribute to accumulating allostatic load in the body, although similar processes also occur internally (mitochondrial allostatic stress, mal). Stress mediators cause structural and functional adaptations of mitochondria – which include the activation of hormonal receptors and the production of reactive oxygen species. If these effects accumulate over time, they can damage mitochondrial DNA – potentially causing mutations or deletions – or induce long-term changes in mitochondrial content and energy-producing capacity. At this point, the mitochondrion begins to produce mitokines that influence cellular functioning by generating allostatic stress biomarkers, increasing oxidative stress, and accelerating cellular aging (Picard, McEwen, 2018). MAL can occur through quantitative processes – such as modification of ATP production parameters and reactive oxygen species – but also through qualitative changes in physiological function – such as alterations in fusion and fission processes or glucose and lipid production. Various aspects of mitochondrial regulation and functioning can influence stress experience in humans. Mitochondria would play a role in cumulative or chronic stress when the accumulation of damage to mitochondrial DNA reaches a limit beyond which energy production is compromised. Additionally, organisms that acquire defects in mitochondrial regulation from experience may exhibit abnormal stress reactivity responses, contributing to raising or inhibiting sensitivity to psychosocial stress. Finally, mitochondria may also provide a protective or buffering mechanism in response to stress by raising antioxidant defenses and cellular resilience. Studies conducted in humans partly support these assumptions, suggesting that early exposure to adverse events – such as maltreatment and abuse in childhood or parental stress – may result in alterations in cellular respiration, mitochondrial DNA copies and integrity, and the presence of free mitochondrial DNA in the blood (Boeck et al., 2016; Ridout, Carpenter, Tyrka, 2016). More recently, Sarah Horn and Phil Fisher conducted a study investigating the association between early adverse experiences, oxidative stress – production of reactive oxygen species – and mental health in a group of 50 adolescents (Horn et al., 2019). The authors documented that adolescents with a history of repeated adverse events had significantly elevated levels of mitochondrial-origin oxidative stress, which in turn were predictive of internalizing symptoms – depressive states or social withdrawal – and the risk of attention deficit and hyperactivity disorder.

As long as life

Telomeres, derived from the Greek words "Telos" meaning "end" and "Meros" meaning "part," are protective caps located at the ends of chromosomes. Comprising repeated sequences in TTAGGG tandem, these sequences can range from a few to over 10,000 bases. Telomeres play a crucial role in maintaining chromosome stability during cell replication and were first elucidated by McClintock in the 1930s. It is now established that telomeres naturally shorten over time and with cell divisions as part of the aging process. The rate of telomere erosion is influenced by genetically determined factors, such as the number of TTAGGG repeats and the activity of enzymes like telomerase, which facilitate their reproduction (Blackburn, 1990). In somatic cells, however, the quantity of telomerase is typically insufficient to consistently preserve telomere length, rendering telomere length an indirect biological measure of cellular aging and a biomarker of the passage of time.

Telomere length is commonly assessed using the method developed by Cawthon (2002), involving separate PCRs to obtain a T measurement of telomeres normalized to the single genetic copy (S). This results in a value known as the T/S ratio (Aviv et al., 2011). At birth, there exists significant individual variability in telomere length, posing challenges for comparative analyses between groups of different subjects, particularly in clinical contexts and cross-sectional study designs that involve a single-point measurement without follow-up. Conversely, longitudinal studies offer the advantage of providing information on the rate of telomere erosion over time in different individuals and may require fewer subjects compared to cross-sectional studies (Aviv et al., 2011).

Physical or psychological environmental stress is one factor that affects the shortening or erosion of telomere length. Studies have observed reduced telomere length in patients with major depressive disorder, bipolar disorder, and schizophrenia (lung et al., 2007; Simon et al., 2006; Wolkowitz et al., 2011; Elvsashagen et al., 2011; Yu et al., 2008). However, some studies have not replicated these results, and the evidence remains inconclusive (Price et al., 2013). Other studies have investigated the role of telomere length in regulating psychosocial stress. For instance, Epel and colleagues (2004) were the first to document an association between psychosocial stress and telomeres in mothers of children with chronic conditions. Subsequent studies found similar results in female caregivers of individuals with chronic and disabling conditions (Parks et al., 2009). Women with a history of domestic violence have shown greater telomere erosion than controls (Humphreys et al., 2012), suggesting that females may be more vulnerable to the effects of telomere stress (Price et al., 2013).

Research also suggests that early stressful experiences can impact telomere regulation. Studies have demonstrated a retrospective association between a history of maltreatment in childhood and reduced telomere length (Tyrka et al., 2010). However, some studies have reported conflicting results (Glass et al., 2010), and the relationship between the number of adverse events and telomere erosion may not be linear (Kananen et al., 2010). Longitudinal studies on preterm babies exposed to high and continuous stress have shown progressive telomere erosion associated with dysregulation of the HPA axis in response to psychosocial stress (Provenzi et al., 2018e, 2019). These findings suggest that early biological learning, reflected in changes in telomere regulation, may influence future stress response systems.

In conclusion, a reliable measurement of telomere length could serve as an index of cellular aging and senescence, as well as a potential biomarker of stress in both children and adults. Prospective investigations with appropriate study methods and designs may reveal whether early supportive, psychological, or therapeutic interventions can slow down or reverse telomere erosion in individuals at risk or exposed to early traumatic experiences, thereby promoting resilience.

Section III

Contexts

9   Waiting for

We do not know when humans first began to imagine a connection between the health of pregnancy and that of the future child and adult. It is possible that this idea slowly emerged through anecdotal observations within different communities and the accumulation of more or less systematic evidence, indicating that individuals raised in less favorable physical and psychological environments tended to show higher rates of medical and psychiatric disorders. Alternatively, the concept might have been proposed in a scientifically testable version in the first half of the thirties of the twentieth century, concerning the observed association between socioeconomic conditions in childhood and life expectancy (Kermack et al., 1934). Subsequently, in the seventies and eighties, further evidence was published regarding the association between living conditions characterized by poverty and the development of heart disorders in adolescence, between pregnancy complications such as preeclampsia and hypertension, and between birth weight and systolic pressure in young adults. In 1986, Barker demonstrated that infant mortality, childhood nutrition, and the risk of heart disease were linked, and he extended these observations to the fetal period (Barker & Osmond, 1986). Since then, the fetal origins hypothesis, more recently identified as the developmental origins of health and disease (DOHaD) to reflect the role of early experiences that can also occur after childbirth, has greatly influenced the development of many lines of research in the field of developmental psychobiology.

At the core of the theories inspired by DOHaD is the concept of developmental programming, which refers to the programming of development by early events or living conditions. It is now almost taken for granted that early environmental conditions can affect the set-point of developmental trajectories of the central nervous system, biology, and behavior, with considerable consequences in the short and long term. Maternal malnutrition during pregnancy, prenatal psychological stress, or dysregulation of the neuroendocrine axis of stress response can send signals to the fetus about the state of the external environment and substantially influence the child's development based on the timing of these signals. Hales and Barker (1992) introduced the concept of the "thrifty phenotype" to refer to adaptations to an environment characterized by limited oxygen and nutrient supply, resulting in reduced fetal growth and altered metabolic efficiency. From Barker's studies,

DOI: 10.4324/9781003479314-12

a series of evidence from animal models and studies conducted in humans have confirmed that a suboptimal intrauterine environment can increase the risk of developing hypertension, insulin resistance, obesity, and type 2 diabetes (Gluckman et al., 2005). Similarly, prenatal maternal stress may be associated with intrauterine growth restriction (IUGR), cognitive deficits, and emotional dysregulation (Talge et al., 2007). Low birth weight may be an indicator of suboptimal intrauterine development and has been used as a predictor of subsequent pathologies in adulthood (Barker, 2004; Whincup et al., 2008). A mechanism that could explain these effects, at least in part, concerns the impact of early adversity in pregnancy on the development of the fetus's hypothalamic-pituitary-adrenal (HPA) axis (Seckl, 2009). However, other processes, such as inflammatory mechanisms and epigenetic regulation of DNA, which have been presented earlier in this book, may also be involved and contribute to a complex interplay that is currently only partially understood.

The placental barrier

Fetal life is a critical period for development and the plasticity inherent in the interaction between genes and environmental disturbances. Recent literature has shown how intrauterine life and environmental signals that reach the fetus can establish mechanisms of adaptation to maternal physiology capable of influencing the future psychological, behavioral, cognitive, and emotional development of the child. Maternal stress during pregnancy, in particular, can disrupt the interaction between the endocrine and immune systems, for example, by causing hyper- or hyposensitivity to cortisol and establishing pro-inflammatory processes involving the action of cytokines. Children born to mothers who experience major depression during pregnancy, with clear alterations in the functioning of their HPA axis, show an increase in cortisol reactivity to stress and, in general, have lower neuro-behavioral development compared to controls (Osborne et al., 2018).

Maternal depressive and anxious symptoms seem to significantly predict higher cortisol levels in children at the ages of 2 and 5 years (de Bruijn, van Bakel et al., 2009) and circadian cortisol regulation patterns at age 15 (O'Donnell et al., 2013; Van den Bergh, Van Calster et al., 2008). These mechanisms may not only increase the risk of pregnancy complications but also lead to alterations in fetal neurodevelopment (Knuesel et al., 2014). An important role is played by the placenta, an organ that acts as a key interface between maternal and fetal biology.

Most maternal cortisol is generally oxidized to cortisone—the inactive form of glucocorticoids—before passing through the placental barrier to fetal circulation. Specifically, cortisol is rendered inactive by a specific placental enzyme—11β-hydroxysteroid dehydrogenase (11βHSD-2)—which acts as a protective barrier, preventing excessive cortisol from the maternal circulation. The regulation of maternal cortisol by these enzymes is crucial during pregnancy, and 11βHSD-2 expression actually changes throughout pregnancy to facilitate different processes that contribute to fetal organ maturation (Howland et al., 2017). The placenta can also synthesize CRH endogenously, with secretions increasing during pregnancy and activating maternal cortisol expression.

Maternal stress related to depressive or anxious states during pregnancy can alter the regulation of 11βHSD-2, thus increasing the risk of exposing the fetus to elevated levels of maternal cortisol (Seckl, 2009). Fluctuations in fetal cortisol exposure may result in dysregulation of the fetal HPA axis, increased sensitivity to stress hormones, and may program the stress response systems of the newborn and infant through alterations in glucocorticoid receptor (GR) expression in the fetal brain (Seckl, 2009). When cortisol levels are excessive, such as during prolonged exposure to stress, depression, or anxiety during pregnancy, they can cross the placental barrier. In rodents, the gene that synthesizes the enzyme 11βHSD-2 exhibits high levels of methylation—gene silencing—in the presence of high levels of chronic prenatal stress and anxiety (Peña et al., 2012). In humans, maternal prenatal anxiety correlates significantly and inversely with the expression of the 11βHSD-2 gene (O'Donnell et al., 2012), while high levels of depression during pregnancy are associated with a high methylation status of the same gene (Conradt et al., 2013).

The gene encoding hippocampal GRs may also be susceptible to epigenetic regulation by maternal prenatal stress. Oberlander and colleagues (2008) replicated the results of Meaney's study, documenting that maternal prenatal depression during the third trimester of pregnancy resulted in an increased methylation level of exon 1F of the NR3C1 gene, at CpG sites similar to those reported in studies on the effects of maternal care in the animal model. In addition, in this study, methylation of the GR gene was positively correlated with salivary cortisol production at 3 months. More recently, Cao-Lei and colleagues examined the prenatal response to maternal stress during the Quebec ice storm in 1998. They documented that maternal stress related to this natural disaster was significantly associated with alterations in cytokine levels (TNF-α and several pro-inflammatory interleukins) in offspring, as well as in the methylation profile of genes involved in the immune system's response to stress (Cao-Lei et al., 2016).

Beyond the barrier

Between 5% and 15% women experience episodes of depression in pregnancy (Biaggi et al., 2016; Giardinelli et al., 2012). This data is even more relevant if we consider that prolonged maternal depressive states can result in a series of complications for the development of the newborn and the child in the first years of life with consequences that may also concern the mental health of the child in adolescence (Gentile et al., 2017; Halligan et al., 2007). Prenatal maternal stress and depression can induce substantial changes to intrauterine growth and may induce early delivery (Grigoriadis et al., 2013; Jarde et al., 2016; Szegda et al., 2014). As seen above, a mechanism contributing to maternal-fetal transmission of prenatal stress involves glucocorticoids and neuroendocrine dysregulation of the HPA axis. However, this is only one of the mechanisms that can be involved. It is in fact possible that a condition of habituation or resistance to glucocorticoids accompanies the condition of prenatal maternal depression and can induce aberrations of inflammatory processes that can also involve the fetus. Glucocorticoid resistance may act by inhibiting the anti-inflammatory action of cortisol on pro-inflammatory cytokines and resulting in an alteration of the function of GRs (Pace et al., 2007; Raison & Miller, 2003).

Elevated levels of pro-inflammatory interleukin IL-6 and TNF-α have been documented in a group of women with a history of prenatal depression (Osborne et al., 2018). In these women, the neuroendocrine response to stress correlated with their baby's cortisol levels, suggesting that the inflammatory response observed in pregnancy may be associated with HPA axis activity. The association between prenatal maternal depression and increased levels of pro-inflammatory cytokines is particularly evident in the third trimester of pregnancy (Osborne et al., 2019). Women with a history of childhood abuse also show higher levels of IL-6 during pregnancy, compared to controls who do not have traumatic events in their childhood (Walsh et al., 2016). It should be noted that childhood trauma is associated with an increased risk of developing depressive disorders in adulthood, even during pregnancy (Plant et al., 2015).

The mechanisms of action of the immune response and its link with maternal depression and the development of the fetus and newborn are poorly understood. It has been hypothesized that the patterns of alteration observed in pro-inflammatory cytokines in depression may alter the production and enzymatic activity acting at the level of the serotonergic system. Cytokines could in fact induce the activity of an enzyme that catalyzes and facilitates the degradation of tryptophan which would not then be transformed into serotonin and would contribute to maintaining high levels of maternal depression (Wirleitner et al., 2003). However, it is highly likely that epigenetic mechanisms are involved in the sensitivity of the fetus to maternal depression. Nemoda and colleagues (2015) recently examined methylome in umbilical cord cells of newborns of depressed mothers and compared it with that of newborns of non-depressed mothers. The researchers noted that different regions were present that had different levels of methylation coinciding with genomic regions related to the production of cytokines involved in the immune response to stress. Pariante's group also showed how prenatal stress can induce changes in microRNA levels in rat brains (Cattaneo et al., 2018).

It is possible that the maternal microbiome may also be involved in the association between maternal stress and long-term programming of the phenotypic profile of the newborn and infant. Carolina De Weerth has shown in macaque infants that prenatal maternal stress can influence the composition of the intestinal microbiome of the baby and that this effect may be particularly significant in the last trimester of pregnancy (Bailey et al., 2004). In turn, the composition of the microbiome has been associated with different stress response and crying profiles (de Weerth et al., 2013; Partty et al., 2012). However, the role of the microbiome in prenatal stress and developmental *programming* is still poorly studied.

On the epigenetic level, it is possible that different genes may act as preferential markers or candidates of *developmental programming* following exposure to maternal prenatal depression. In addition to the aforementioned study by Oberlander et al. (2008), more recent research has confirmed a key role played by the regulation of methylation of the gene encoding GRs: maternal depression has been associated with an increase in *NR3C1* gene methylation in children at 2 (Braithwaite et al., 2015) and 14 months of age (Murgatroyd et al., 2015). Importantly, the evolutionary outcomes of an increase in methylation of *NR3C1* include lower neurobehavioral (Conradt et al., 2013) regulation, greater neuroendocrine (Oberlander et al., 2008) and behavioral (Sheinkopf et al., 2016) reactivity to stress, and reduced attentional abilities (Bromer et al., 2013). A systematic review of the literature has highlighted how such effects can be particularly significant if exposure to stress—as in the case of prenatal maternal depression—occurs in the first thousand days after conception (Berretta et al., 2021). The gene encoding the serotonin transporter (*SLC6A4*) is susceptible to changes in epigenetic regulation in response to environmental stimulation and an increase in methylation levels of this gene appear to be a reliable biomarker of early stress (Provenzi et al., 2016a). Oberlander's group reported altered methylation levels of *SLC6A4* in children exposed to maternal depression (Devlin et al., 2010), although a direct link between methylation and expression of this gene is not obvious and the role played by the 5-HTTLPR polymorphism cannot be excluded (Provenzi et al., 2016; Wankerl et al., 2014). Eva Unternaehrer and colleagues (2016) reported alterations in methylation levels in different regions of the gene encoding the oxytocin receptor—*OXTR*—in subjects exposed to prenatal depression.

Studies conducted on large portions of the genome and not on single target genes have identified a variable number of CpG sites that may exhibit variations in methylation levels in association with the degree of severity of maternal depressive symptoms in pregnancy (Nemoda et al., 2015; Non et al., 2014, Schroeder et al., 2012). However, it should be noted that in epigenome-wide studies the associations with the degree of methylation of the identified CpG sites may be partially informative, since they may concern portions of genes whose association with maternal stressors or with developmental outcomes in the child are only partially known (Provenzi et al., 2018a). The study of the epigenetic mechanisms involved in the alteration of the immune and neuroendocrine response to maternal depression is only at the beginning. However, it is reasonable to expect that research in this area will provide important revelations regarding how early adverse experiences that can occur well before birth can be incorporated into the behavioral profile of the child through epigenetic modifications (Provenzi et al., 2017a).

Caregiving programming

The developmental programming paradigm is contributing enormously to provide a meta-theoretical framework for understanding how early life experiences can set different levels of risk for physical and psychiatric illnesses. Central to this approach is the idea that there is a typical line of development with respect to which it is possible to identify deficits or deviations of evolutionary trajectories that can be defined as atypical or aberrant (Toth & Cicchetti, 2010). As we will see later—in Chapter 13 of this volume—this is at least a partial approach, which can have important repercussions and limitations for clinical practice. However, although the fetal period has been characterized as a time window of high neuroplasticity for the future newborn and child, less attention has been paid to studying how pregnancy can be an equally flexible period of life for the woman.

This is surprising if we consider the fact that pregnancy involves a series of changes in the physiology, biology, and behavior of women. Among these changes, the most evident is perhaps the growth and development of a new organ—the placenta—on which we have already focused, highlighting its neuroendocrine and immunological impact on the maternal-fetal compartment. Laura Glynn in California and Ruth Feldman in Israel are among the researchers who have been more extensively concerned with understanding what are and how the mechanisms of maternal neuroplasticity occur during pregnancy from the perspective of the maternal organism. As a result of physiological changes that accompany pregnancy, childbirth, breastfeeding, and interaction with the newborn, women show changes that involve behavioral and emotional regulation, but potentially also brain and cognitive functioning. In animal models, it is known that maternal reproductive history is related to brain structure, as well as to functional responses and neurogenesis (Brummelte et al., 2006; Byrnes et al., 2013). These modifications, in rodents, can be maintained for several years and can accumulate with the increase in the number of litters.

In humans and primates, as well as in rodents, prenatal neuroendocrine regulation facilitates the emergence of parental sensitivity behaviors in postpartum females (Feldman et al., 2007). Hormonal exposure during pregnancy also facilitates the bonding of maternal attachment with behaviors that can be observed even until the end of the first year (Glynn et al., 2016). Cognitive effects have received less attention in the literature and present discrepancies between the animal model and humans. In rodents, greater cognitive ability has been observed during pregnancy: better spatial and working memory, greater motor planning skills, reduced latency of response to the environment (Pawlusky et al., 2006). However, in women, attentional functions and memory efficiency do not have the same advantages and can even be reduced. As we have already seen, in the absence of prolonged states of stress, depression, or anxiety, cortisol production in response to acute stressful events during pregnancy is progressively reduced as we approach childbirth (Buss et al., 2009; Glynn et al., 2018). However, cortisol is only one element of the puzzle. From a cerebral point of view, motherhood is associated with an increase in functional connectivity in regions involved in sensitive parental behavior and maternal responsiveness to child signals (Atzil et al., 2011; Swain et al., 2017). Hoekzema and colleagues (2017) recently showed how changes observed in gray matter in pregnant women may be associated with maternal attachment and can last up to two years after delivery.

The fetus can also exert more or less direct influences on the physiology of the pregnant woman through at least three different languages: hormonal regulation, its movement, and genetic exchange. As we have already mentioned, the placenta contains mostly cells of fetal origin—trophoblasts—capable of producing hormones that enter the maternal circulation (Cuffe et al., 2017). It is still unclear how these hormones contribute to the maternal phenotype. However, some evidence suggests that the oxytocin produced by the fetus increases during gestation, and mothers who show high levels of oxytocin during pregnancy report feeling more emotionally attached to their baby, are better able to interpret its communicative signals and are more frequently involved in eye exchanges, contacts, and affective tactile stimulation (Feldman et al., 2007). Moreover, fetal movements can be implicitly perceived by maternal physiology, with an increase in sympathetic activity to which the mother does not seem to get used during gestation (DiPietro et al., 2015). It has been hypothesized that these physiological responses can function as precursors and preparers of the behavioral synchrony that can be observed after birth, thus initiating a dyadic exchange that will then take the form of a real interactive dance (Glynn et al., 2018). Finally, since the seventies, it is known that some fetal cells can be traced in maternal blood even many years after childbirth (Herzenberg et al., 1979). It is a phenomenon known as fetal chimerism, and it is still unclear whether it can contribute to maternal health or instead have adverse effects. On the one hand, in fact, it seems that these cells may contribute to facilitating the response of maternal biology to wounds, while other studies associate the presence of fetal cells in the maternal circulation with pregnancy complications (Gammill et al., 2013).

Prenatal conflicts

Until the seventies, maternal-fetal biology was generally conceived as an interaction characterized by synergy, cooperation, and harmony. However, Trivers (1974) was the first to suggest that the relationship between mother and fetus can instead be better understood as a conflict between different evolutionary strategies and by a sort of conflict generated by different physiological and biological interests. To better understand the nature of the bidirectional biological relationships between mother and fetus, it is possible to conceive that hormones that come from the maternal to the fetal compartment may actually be informative messages about the state of the extrauterine environment and therefore can somehow help the fetus to prepare for postnatal life. It is an evolutionary assumption that the fetus could be facilitated by making predictions about the external environment, starting to adapt its physiology and biology to maximize survival (West-Eberhard, 2003).

According to this perspective, it is possible to understand some of the adverse consequences of prenatal stress as adaptive responses of the fetus or as side effects of an attempt to make predictions based on indirect information mediated by maternal biology. The offspring of mothers presenting with prenatal stress states more frequently show anxious behaviors, reduced attentional capacity, and basal and reactive dysregulation of the HPA axis (van den Bergh et al., 2008; Cirulli et al., 2009). It is possible to consider the physiological and biological signals of maternal stress that reach the fetus as informants, parameters that allow identifying some elements of the environment that can influence adaptive strategies and reproductive success. In other words, it is possible to imagine that high levels of stress hormones are signs of a potentially harmful or dangerous environment (Kapoor et al., 2006). The fetus could, therefore, prepare to respond to a hypothetical environment characterized by high stress by modifying its neuroendocrine and behavioral structure accordingly. Obviously, these are changes that have a cost and that can prepare for the future environment only on a probabilistic basis and that—especially nowadays, in which the cultural aspect and technical innovation have enormously modified the external environment in which we live—can be fallacious or only partially functional. In humans, exposure to prenatal stress may be associated with phenotypic profiles characterized by high anxiety and fear, dysregulation of behavior and response to psychosocial stress, attentional and peer interaction difficulties (Glover et al., 2009; Pluess & Belsky, 2011). However, developing high stress responsiveness can also be beneficial if the individual lives in contexts where this trait can confer adaptive benefits (Boyce & Ellis, 2005).

The perspective that sees in maternal-fetal biological exchanges a language or an adaptive communication system on environmental conditions generally considers that the interests of the fetus—and of the future newborn—are largely favored by letting maternal biology take its course. However, this interpretation is partial because it neglects to consider the fact that the mother can use her biological devices to promote her well-being, sometimes even at the expense of the developmental programming of the fetus. Del Giudice (2012) proposed that, in fact, biological prenatal communication between mother and fetus can be better understood in terms of a potential or real conflict between the evolutionary and survival or adaptation interests of the two interactive "partners." As anticipated, Trivers (1974) had already proposed a theory of maternal-infant conflict, and a version of this theory indirectly influenced the development of Belsky's theories concerning individual variation in differential susceptibility to stress. Applying this theory to the maternal-fetal compartment, Del Giudice (2012) proposes that during pregnancy, to optimize its biological adaptation, the fetus tries to extract as much information (e.g., nutrients) from the mother even if this can be dangerous for maternal health, while the mother tries to counterbalance by reducing or altering the same flow in the opposite direction.

In the nineties, Haig (1993) analyzed maternal-fetal biological interactions from the point of view of the theoretical perspective of Trivers (1974) succeeding in shedding light on a series of evidence that—with a view to a mother-fetal syntonic collaboration—were difficult to interpret. For example, in mothers, insulin levels rise significantly during the last trimester of pregnancy, but at the same time, women develop insulin resistance. This can have important negative repercussions on the woman's health by increasing the glucose content in the blood. This is a paradox that makes sense only if it is assumed that the increase in endogenous insulin resistance is stimulated by the fetus through the secretion of hormones (e.g., growth hormone) via the placenta (Haig, 2007). Maternal-fetal regulation of blood pressure can also be another example. Blood pressure rises during pregnancy and remains elevated in the third trimester. At the same time, vasodilation decreases arterial resistance in the first two trimesters. The effect is a dizzying drop in blood pressure about mid-gestation. Vasodilation, like insulin resistance, also seems an unnecessary challenge to the physiology of pregnant women. However, it can be understood as a defensive response of maternal biology to the fetus-induced increase in vasoconstriction (Haig, 2007).

These examples suggest that the fetus is an active participant in maternal-fetal life and not just a passive recipient of one-way communication. The conflict between the biological interests of mother and fetus can result in the emergence of counterintuitive and complex interweaving of the physiological regulations of both. Most of the strategies of armament and conflict resolution reside in the production of hormones that enter the maternal or fetal cycle and that see the placenta as a crucial organ (Haig, 1996). One of the reasons why this conflict could have evolutionary reasons is related to the role played by stress—even prenatal stress—in favoring processes of neuronal plasticity and epigenetics. Boyce and Ellis (2005) identified this plasticity as a biological sensitivity to the environment, and Belsky developed the concept of differential susceptibility (Belsky, 1997). In species characterized by prolonged postnatal maternage, the mother becomes a decisive part of the postnatal environment. According to the model of Del Giudice (2012), the mother would benefit from being able, in some way, to increase the susceptibility of her fetus and newborn to environmental influences, that is, to those of her own caring behavior. At the same time, the fetus would try to maximize resistance to thrusts that increase its plasticity (Belsky & Pluess, 2009). However, if early exposure to stress in pregnancy is a crucial element in defining neuroplasticity gradients, mother and fetus face conflict because they have different interests but are still willing to reach the best possible compromise between plasticity and adaptation, between complexity and coherence of the system. It is probably no coincidence that the same genes that contribute to the regulation of stress—such as genes involved in the functioning of the serotonergic and dopaminergic systems—are also the same genes involved in neuronal plasticity in the animal model, in the rodent, and in the macaque (Veenema et al., 2004). Since prenatal stress may be able to contribute to the long-term programming of the HPA axis and behavioral reactivity, postnatal plasticity can be similarly programmed by prenatal adverse events, especially in individuals who are biologically susceptible or who have biological sensitivity to the context.

From the perspective of the fetus, a potential defense mechanism against stress, arising from biological signals such as inflammation or elevated cortisol levels resulting from prenatal depression, involves the partial filtration of hormones through enzymatic barriers at the placental level. Simultaneously, maternal biology may increase cortisol production or interfere with the placental barrier, which is less biologically costly. The fetus may also activate mechanisms that reduce the hormonal load at the placental level, utilizing other hormones like oxytocin or placental-produced anti-inflammatory agents. Generally, if the biological signals are optimal for the mother, the fetus will experience reduced activation of plastic responses, resulting in a phenotype more aligned with maternal biology. Conversely, if the fetus can effectively activate the filtering process undisturbed, the phenotype will be better suited to fetal adaptation. The presence of side effects from conflicting strategies or tactics may lead to phenotypic alterations, which can be seen as long-term programming changes or early biological learning.

10 Born soon

Preterm birth, accounting for around 5% to 18% of global births, is a significant risk factor for perinatal health. Babies born preterm, before 37 weeks of gestation, face an increased risk of neurobehavioral disorders, sensory issues, behavioral problems, and psychomotor and cognitive retardation. Even in the absence of serious pathologies or perinatal damage, preterm infants are exceptionally vulnerable and often require hospitalization in neonatal intensive care units (NICUs) for weeks or months.

The establishment of intensive care wards for at-risk newborns is a relatively recent development. In the 1960s, American neonatologist Marshall Klaus pioneered the opening of neonatal units for premature babies in the presence of parents. His works, "Care of the High-Risk Newborn" and "Mothering the Mother," have long been considered essential references for neonatal intensive care specialists and have significantly contributed to the evolution of family-centered early care. Concurrently, Barry Brazelton and Heidelise Als played vital roles in making baby observation a fundamental practice in perinatal care. Brazelton's scale enabled the assessment of newborn neurobehavioral profiles and facilitated targeted preventive and therapeutic interventions based on individual strengths and vulnerabilities. Heidelise Als focused her research specifically on preterm babies and newborns at risk. Her work on the neurophysiology of newborns has provided insights into how early changes in the care environment in the NICU can have lasting effects on the baby's development through central nervous system alterations. Among the outcomes of her extensive work, the Newborn Individualized Developmental Care and Assessment Program (NIDCAP) has internationally promoted a culture of family-centered care in neonatal intensive care. Notably, Als emphasized the significance of recognizing the child's signals and behavioral cues, as well as their capacity to regulate themselves. The sinactive theory of infant development, developed by Als in the 1980s, has further provided a meta-theoretical framework for interpreting the behavior of at-risk infants. The concept of kangaroo care emerged in Bogotà, Colombia, in the late 1970s when Rey and Martinez, facing limited technical resources and high infection risks, introduced continuous skin-to-skin contact on the mother's chest to maintain newborns' thermal regulation. Recent studies have demonstrated that kangaroo care can be more effective than incubators, reducing the neonatal mortality rate by about 50% in stable preterm babies.

DOI: 10.4324/9781003479314-13

However, certain aspects of the NICU environment can act as stressors for pre-term infants, such as painful stimuli, disrupted sleep, excessive noise and light stimulation, frequent handling, and separation from parents (Peng et al., 2009). While some of these stimulations are necessary for the child's survival, their adverse effects can be mitigated by maximizing parental involvement in the NICU. Parental engagement serves as a primary prevention strategy, aiming to enhance the well-being and stability of both the child and the parents. This approach should be adaptable to the specific needs and resources of each family during the hospitalization period. Various practices involving parents in the NICU are collectively known as developmental care (Westrup, 2007). Developmental care encompasses a range of strategies and interventions designed to minimize the impact of stressful NICU procedures. This includes controlling the intensity and volume of invasive stimuli, facilitating early parent-child contact and parental involvement, supporting the child's neurobehavioral development, and managing neonatal pain. Over the past three decades, parents have increasingly become active participants in NICU care, although significant variations persist across countries and specific national contexts (Montirosso et al., 2017).

This chapter aims to provide an introduction to the context of preterm birth and early NICU hospitalization as early risk factors for the socio-emotional and socio-cognitive development of the child, as well as the well-being of the parents. We will present evidence suggesting that the early stress experienced during NICU hospitalization can lead to lasting dysregulation of physiological and neurobiological processes. Particular attention will be given to recent findings in behavioral epigenetics of prematurity (Provenzi et al., 2018d), which explore the epigenetic mechanisms underlying the adverse effects of NICU experiences (Provenzi, Borgatti, and Montirosso, 2016a). Conversely, psychobiological studies on developmental care procedures demonstrate that the same mechanisms can contribute to the neuroprotective effects of promoting parental involvement in the NICU.

Dark side

Children born too early are then admitted to the NICU, an environment in which physical proximity to parents is often limited or partial, and in which they are exposed to high levels of physical, sensory, and painful stress. It should be emphasized that preterm infants have an immature central nervous system and organs responsible for the stress response, especially when considering that the external environment is vastly different from the intrauterine one. Therefore, environmental stimulations during this delicate postnatal period can be particularly dangerous and can have lasting effects on developmental programming.

For one thing, children in the NICU are exposed to physical and sensory stimulation characterized by high levels of intensity and duration, which are difficult to tolerate for preterm infants (Brown, 2009; Ozawa et al., 2010). Prolonged exposure to artificial lights—which in some cases may exceed the limit of 600 lux (Lee et al., 2005)—and noises—often above the recommended threshold of 45 decibels (Altuncu et al., 2009)—can be extremely disturbing, with negative consequences for the long-term development of the sensory system and the stabilization of the sleep-wake rhythm (Calciolari & Montirosso, 2011; Graven, 2004).

In addition, clinical procedures such as intubations, punctures, sampling, and surgery—also manipulations and removal of patches—can be particularly painful for preterm infants. Given their condition of sensorineural immaturity, premature babies have low thresholds of resistance and protection to pain. Invasive and painful procedures—generally referred to as skin-breaking—have been extensively studied by Ruth Grunau in preterm babies (2013). These procedures may result in short- and long-term alterations in hypothalamic-pituitary-adrenal (HPA) axis regulation, including circadian rhythm dysregulation (Brummelte et al., 2011), and programming of a neuroendocrine hypo-responsiveness to stress profile (Haley et al., 2006; Provenzi et al., 2016d). Further consequences may include structural and functional alterations of the central nervous system that can persist into subsequent years (Smith et al., 2011) and result in cognitive deficits (Zwicker et al., 2013), as well as emotional and behavioral disorders (Ranger & Grunau, 2014).

It should also be noted that exposure to these sources of early stress often occurs in the absence of adequate promotion of supportive contact and the regulatory component implicit in parent-child contact. The forced separation between parents and child in NICU can interrupt and disturb the emergence of an early affective bond (Latva et al., 2004). In the animal model, early maternal separation has observable effects both on the development of the central nervous system (Bock et al., 2014) and on the programming of the neuroendocrine response axis to stress (Welberg & Seckl, 2001). In humans, early alteration of the parental care context can alter and reduce early physical contact and make the child more exposed to the neurotoxic consequences of neonatal pain (Morelius et al., 2007).

Finally, we should not underestimate the stressful impact that hospitalization in NICU can also exert on the well-being of parents (Montirosso et al., 2012; Provenzi & Santoro, 2015). On the one hand, the mothers and fathers of preterm babies suddenly find themselves immersed in a highly technological, cold context, where their child—born earlier than expected—is connected to pipes and machinery that, in many cases, keep him alive and separate him from contact with the parent. These parents may develop emotional disorders, depression, anxiety, and be hindered in the process of developing an affective bond of attachment to their child (Miles et al., 2007; Montirosso et al., 2014; Provenzi et al., 2017c).

Protecting the newborn and the parents

As anticipated, the practices of developmental care include a series of interventions that aim to promote and support direct parental involvement in NICU, reducing, containing, and mitigating the effects of stress on the child and the parents themselves (Als et al., 2004). It should be emphasized that the practices of developmental care are not uniformly applied in NICUs (Ashbaugh et al., 1999; Greisen et al., 2009). In an Italian multicenter study, Montirosso and colleagues (2012) documented wide variability in parental involvement in NICU and showed how children who come from NICUs that promote more parent-child presence and early contact have more adaptive developmental outcomes, in terms of behavior, cognitive abilities, and quality of life (Montirosso et al., 2017). NICU children for whom parental presence is infrequent and inadequately encouraged are at a higher risk of developing psychological problems over the years (Latva et al., 2007).

Among the most used developmental care strategies are skin-to-skin contact (Feldman et al., 2002) and neonatal pain management (Pillai Ridell et al., 2011). Skin-to-skin contact can have benefits not only for the promotion of the physiological stability of the child before discharge (Cong et al., 2011), but also on the organization of sleep (Calciolari & Montirosso, 2011), brain maturation (Korraa, El Nagger et al., 2014; Scher et al., 2009), and behavioral and emotional development (Keren et al., 2003). At the physiological level, children continuously exposed to skin-to-skin contact with the parent show reduced reactivity to acute stressful stimulation at the age of 1 month (Morelius et al., 2015). In the long term, the adoption of parental involvement strategies in NICU is associated with better psychomotor development at the age of 2 months (Kiechl-Kohlendorer et al., 2015). In addition, promoting knowledge of one's child through an early video-feedback intervention and support for parenting in NICU can facilitate the development of psychomotor skills, particularly in the areas of sensory and socio-relational regulation (Pisoni et al., 2021).

Developmental care practices can also have beneficial effects on parental well-being (Matricardi et al., 2013). Supporting parents to engage in early physical contact with their preterm infant and facilitating breastfeeding are interventions that can support the transition to parenthood and increase the sense of affective attachment to the newborn (Flacking et al., 2006). In a recent study conducted by Martha Welch's group (Welch et al., 2020), preterm mother-infant dyads were randomly assigned to two groups. One group received the normal care required by the NICU protocol, while the other received a standardized intervention aimed at increasing physical and emotional contact between mother and newborn—the Family Nurture Intervention (Hane et al., 2019; Welch et al., 2013). In this intervention, the main objective is not so much to promote skin-to-skin contact practices, but the emotional connection between mother and child. The intervention includes a series of sessions of about an hour, called calming sessions, in which mothers are encouraged to express their feelings and speak emotionally to their child. Initially, these sessions take place while the baby is still in the incubator; however, subsequent sessions can also be carried out outside the incubator, in a more comfortable way for the parent. This intervention has been widely used by Martha Welch's group: its effects include an improvement in maternal sensitivity (Hane et al., 2015), a reduction of symptoms of anxiety and depression in mothers at 4 months of corrected age (Welch et al., 2016), greater brain maturation of the child (Myers et al., 2015; Welch et al., 2017), better psychomotor development, and a lower incidence of attentional problems in children at 18 months (Welch et al., 2015). In the work published by Welch and colleagues in 2020, children and mothers in the clinical group who had participated in Family Nurture Intervention sessions had better parasympathetic regulation, as shown by higher levels of RSA at rest and a more rapid increase in vagal tone in response to maternal contact than the control group.

In essence, it is possible to observe how stressful environmental exposures present in NICU can have harmful effects on the programming of stress reactivity (Provenzi et al., 2016) and on neurobehavioral development (Lester et al., 2011), partly by altering brain growth and activity (Ranger et al., 2015; Smith et al., 2011), resulting in an increased risk of emotional-behavioral disorders in preschool and school age (Chau et al., 2014). At the same time, developmental care can exert neuroprotective effects of the opposite direction: facilitating neuroendocrine regulation of stress (Kleberg et al., 2008), neurophysiological stability of the newborn (Montirosso et al., 2012), behavioral regulation (Montirosso et al., 2018), and adequate growth and complexity of the central nervous system (Als et al., 2004). We can then ask ourselves what are the mechanisms—the neurobiological pathways—through which the early experience of hospitalization in NICU can exert these effects on the preterm infant. This question is the basis of the development of a line of studies that is proving particularly interesting both on a scientific level and on that of clinical implications and support policies for early care: the behavioral epigenetics of prematurity (Provenzi, Guida, and Montirosso, 2020).

Music power

Although promoting early contact is one of the most widely adopted strategies to support the neurobehavioral development of preterm infants and enhance parental well-being, maternal voice can also be an effective tool (Filippa et al., 2017). Interventions using maternal voice in the NICU make use of pre-recorded sound stimuli or ask parents to speak or sing near the incubator (Provenzi et al., 2018b). Rand and Lahav (2014a) suggested that maternal voice can offer the preterm neonate a unique opportunity for sensory enrichment starting from 26 weeks post-conceptional age, especially if physical contact presents implementation challenges due to the child's neurobehavioral instability.

A recent literature review (Provenzi et al., 2018b) collected evidence from 18 original studies conducted on more than 300 mother-infant dyads. In these studies, the effect of maternal voice was tested in groups of children born very preterm (Chorna et al., 2014; Picciolini et al., 2014)—below 32 weeks of gestational age—as well as in children born mildly preterm (Butler et al., 2014). The systematic exposure to maternal voice occurred between 23 and 36 weeks of post-conceptional age. Most original studies used exposure to pre-recorded maternal voice to rigorously control sound characteristics and stimulation intensity. However, other researchers studied the effect of live maternal voice while the parent was near the incubator (Caskey et al., 2014; Filippa et al., 2017). Maternal voice could also be mixed with sounds resembling the heartbeat to compensate for the loss of the intrauterine acoustic environment (Rand & Lahav, 2014b). Mothers could talk, read stories (Filippa et al., 2017; Panagiotidis & Lahav, 2010; Saito et al., 2009), or sing (Cevasco, 2008). Controlled exposure to maternal voice differed in various studies: it could last from a few minutes (Saito et al., 2009) to several hours (Webb et al., 2015) and be repeated for a variable number of sessions.

Children exposed to maternal voice showed alterations in autonomic regulation, with an increase in heart rate observed in some cases (Filippa et al., 2017) and a decrease in others (Picciolini et al., 2014; Rand & Lahav, 2014b). Listening to maternal voice might increase oxygen saturation (Filippa et al., 2013), and the number of critical cardiorespiratory events decreased during maternal voice exposure (Filippa et al., 2017). Saito and colleagues (2009) documented effects at the brain level using near-infrared spectroscopy (NIRS), highlighting specific activations in response to maternal voice or that of a nurse in the left and right frontal cortex. In a later study, infants exposed to maternal voice showed greater amplitude of bilateral auditory cortex at 1 month after birth (Webb et al., 2015). Children were more capable of maintaining wakefulness during maternal voice exposure (Filippa et al., 2013). Furthermore, an Italian study showed better visual and auditory orientation skills at 3 corrected months of age in infants exposed to an intervention involving exposure to their mother's voice (Picciolini et al., 2014). However, these differences were no longer found at 6 months. The length of hospitalization could be reduced—on average by two days—in children who were extensively exposed to pre-recorded maternal voice (Cevasco, 2008). Maternal voice in the NICU also seems to be associated with the child's greater ability to engage in proto-conversational turns after discharge and higher scores in cognitive and linguistic development between 7 and 18 months (Caskey et al., 2014).

Maternal voice, therefore, seems to partially contribute to the positive effects observed in response to developmental care interventions in preterm infants. Although the evidence is not conclusive, and there remain considerable differences in various studies, mostly related to the different characteristics of the tested children and the NICU environments (Provenzi et al., 2018b), it is possible to hypothesize that allowing parents to engage with their voice in early interactions during the NICU stay may be a particularly effective intervention. Maternal voice is generally a non-harmful and noninvasive stimulus for the infant and—in environments where NICU equipment noise is adequately controlled—it can enable an early encounter between parents and the newborn, contributing to the behavioral, relational, and cognitive developmental trajectory of the child.

Epigenetics and neuroplasticity

As mentioned in Chapter 7 of this volume, DNA methylation is a highly susceptible epigenetic regulation process to environmental variations, which remains at least partially stable throughout life, contributing to the individual phenotype. DNA methylation is susceptible to both adverse events (Griffiths & Hunter, 2014) and variations—for better or worse—in parenting behavior (Weaver et al., 2004). Therefore, it is not surprising that the role of epigenetics during the early postnatal period may be particularly relevant for the processes that determine the neuroplasticity of the central nervous system (Fagiolini et al., 2009).

Two biological systems are of particular relevance to understanding the relationship between stress regulation, epigenetics, and neuroplasticity: the HPA axis and the serotonergic system. The HPA system has already been extensively discussed in Chapter 5 of this volume: it is the main neuroendocrine system regulating stress, develops during the first months of life, and reaches maturity only after the first four years. It works through a hormonal cascade reaction that results in the production of cortisol and is capable of self-regulation through feedback mechanisms that exploit specific receptors (GR) at the hippocampal level.

The serotonergic system plays a critical role in the development of the HPA axis and contributes to the modulation of its functioning (Andrews & Matthews, 2004). Serotonin is a neurotransmitter widely distributed in the central nervous system. It is released in response to stressful events, and its activity in the inter-synaptic space is regulated by the serotonin transporter (5-HTT). This molecule removes serotonin from the synaptic space and replaces it in the presynaptic button, resulting in increased serotonergic turnover. Therefore, any change in the signaling and activity of the serotonin transporter can lead to dysregulation of the serotonergic system and may contribute to alterations in the functioning of the neuroendocrine system. In fact, within the animal model, serotonin is able to contribute to the expression of GR receptors in the hippocampus in a contingent way with the exposure of the offspring to physical contact and stimulation by the parent.

As already widely shown in Chapter 7, the gene-encoding cortisol GRs in humans—the NR3C1 gene—and the gene that contains instructions for 5-HTT synthesis—the SLC6A4 gene—are both susceptible to epigenetic variations in response to parental stimulation. It is, therefore, not surprising that these genes have been particularly investigated in the model of the preterm born, to understand how behavioral epigenetics can contribute, at least in part, to explain the effect of early environmental exposures that, for better or for worse, reach the premature baby and alter its developmental trajectories.

Behavioral epigenetics of prematurity

In 2015, Montirosso and Provenzi proposed a coherent model for applying the lens of behavioral epigenetics to the study of preterm infant development. In this model, both stressful and neuroprotective exposures exert an influence on the child's future phenotype through epigenetic processes, particularly alterations in the methylation levels of genes involved in behavioral regulation, stress response, and the development of socio-emotional and socio-cognitive skills (Provenzi & Barello, 2015; Provenzi & Montirosso, 2015).

The first study in this line of research was conducted by Kantake and colleagues (2014). In this study, the methylation of several CpG sites of the NR3C1 gene increased from birth to the fourth postnatal day in preterm infants, while it remained relatively stable during the same period in term-born infants. Regression analyses revealed that the methylation levels observed on day four were significantly associated with gestational age and low birth weight. Moreover, higher levels of methylation corresponded to infants with more clinical complications, such as lung problems or cardiovascular instability. In a retrospective study in the same year, methylation of the SLC6A4 gene was measured in salivary samples of seven-year-old preterm infants. Children who had been exposed to a greater number of skin-breaking procedures during NICU stay showed higher levels of methylation of the gene encoding for 5-HTT, which were significantly associated with the risk of developing emotional and behavioral problems at seven years.

In a longitudinal study, Provenzi et al. followed preterm infants born before 32 weeks of gestation and term-born infants from birth to five years of age. They measured the methylation levels of the SLC6A4 gene at birth—from cord blood—and at discharge from the NICU—from peripheral blood, but only in preterm infants—and obtained a series of measures of socio-emotional, cognitive, and brain development in the subsequent years. In this study, the methylation at birth of the gene that synthesizes 5-HTT was not different between preterm and term-born infants. However, as the number of painful stimulations—skin-breaking procedures —increased, the level of methylation increased in specific CpG sites near the gene's promoter region. High levels of methylation in these CpG sites were predictive of modifications in parietal brain volume observed at discharge. Moreover, the reduced brain volume in these areas was predictive of the children's cognitive development at 12 months. At 3 months of age, preterm infants who had undergone more painful interventions and showed an increase in methylation of the SLC6A4 gene were rated by their parents as less able to pay attention and less available in social interaction. At the same age, the negative emotional response to the still-face procedure was not only higher in the preterm group compared to the term-born group, but the difference was partially explained by the high levels of methylation observed in one of the CpG sites highlighted in previous works.

In another study, Provenzi et al. (2017d) investigated whether maternal sensitivity during the play phase of the still-face paradigm could play a buffering and protective role with respect to the stress experienced during the still-face. In the group of term-born infants, higher levels of maternal sensitivity reduced the levels of negative emotion observed in the child, regardless of the methylation levels observed at birth in the SLC6A4 gene. On the contrary, in the preterm group, there was no moderating effect of maternal sensitivity, suggesting that parenting support interventions should be adequately initiated to help parents provide effective support for stress regulation strategies in the early months of life after discharge from the NICU.

At the age of four and a half years, the same children participated in a standardized frustration and emotional regulation laboratory test, the Preschooler Regulation of Emotional Stress (PRES; Provenzi et al., 2017b). In this test, the child's response to four conditions in succession was observed: interaction with a stranger, the request to draw a perfect circle, the delay of an expectation, and the request to open a transparent box closed with a wrong key. In the preterm group, the level of methylation of SLC6A4 observed at discharge predicted the stress response—manifested as anger responses—to this procedure, while no significant associations were found in the term-born group. These data suggest that epigenetic changes in the gene encoding for the serotonin transporter can be more influential on the child's subsequent development, the more they occur in response to environmental variations.

In the same longitudinal project, Provenzi et al. also measured telomere length from the same blood samples obtained at birth and at discharge from the NICU. In the group of preterm infants, higher levels of neonatal pain were associated with increased telomere erosion observed at discharge. In particular, children exposed to frequent pain showed a reduction in telomere length, while children exposed to low levels of pain showed resistance to cellular aging. At 3 months, telomere erosion observed between birth and discharge from the NICU was able to predict the neuroendocrine response to the still-face procedure: children who showed shorter telomeres responded to the maternal still-face with a decrease in cortisol levels. These results suggest that the development of a pattern of HPA axis hypo-responsiveness to stress may be, at least partially, influenced by epigenetic processes that occur at the telomeres' level.

Light side

Until now, only a few studies have focused on the implications of epigenetic processes regarding the neurobiological effects of developmental care procedures. Only in 2020, the group of Katrin Mehler in Germany published a study reporting an association between prolonged skin-to-skin contact exposure and alteration in the expression of genes involved in stress regulation (Hucklenbruch-Rother et al., 2020). In this study, 88 preterm infants born before 32 weeks of gestational age were assigned—after achieving physiological stability—to either 60 min of skin-to-skin contact with the mother or 5 min of visual exposure to the mother without physical contact. The RNA expression of six genes, including NR3C1 and SLC6A4, was measured using peripheral blood samples obtained when the infants reached 40 weeks of gestational age. The expression of NR3C1, SLC6A4, and another gene involved in corticotropin release, CRH R2, was higher in the group of infants exposed to 60 min of skin-to-skin contact. Furthermore, CRH R2 expression levels predicted the neuroendocrine stress response at the corrected age of 6 months. While this is the only study that has investigated the effect of parent-child relationship support interventions in the NICU, the results are promising and suggest that developmental care interventions may also exert short- and long-term effects through the same epigenetic processes involved in neonatal pain and stress. Indirect confirmations come from a study conducted by Barry Lester and Ed Tronick (Lester et al., 2018) in which exposure to maternal breastfeeding in the first 5 months was significantly associated with lower methylation of the NR3C1 gene and reduced HPA axis reactivity to stress elicited through still-face procedures at 5 months of age.

The study of epigenetic processes related to stressful exposures and developmental care interventions to which preterm infants are exposed appears to be a highly promising line of research. From a scientific perspective, these studies can help us understand how variations, for better or worse, in early caregiving conditions can have long-term effects on the child's well-being. The preterm infant, in fact, allows us to conduct studies of behavioral epigenetics at a very early age, corresponding to the fetal period but outside the uterine environment. At the same time, from a clinical perspective, the implications of these studies can provide important insights into the consequences of stressful exposures in the neonatal unit. A thorough understanding of the biological changes associated with neonatal practices in the NICU can help us improve prevention and care actions for preterm infants and their parents, contributing to enhancing developmental care programs.

11 Embodied parenting

The history of art is rich in depictions of the relationship between parents and children. However, these are often maternal-infantile images: Picasso's Mother and Child (1921), Gauguin's Orana Maria (1891), Raphael's Madonna of the Meadows (1508), Renoir's Child with Toys (1895), Klimt's Three Ages of Woman (1905), to name a few. Classically, when depicted, the father appears with characteristics very different from those typically associated with the maternal figure. For example, the father shows the child to the world in the famous gesture of Hector; the depiction of Silenus with Dionysus, from 300 BC, is one of the rarest examples of paternal representation in intimacy with his own child. Paradoxically, in much of Disney's artistic production, the maternal figure is often absent, and some paternal figures take the stage: Jasmine's sovereign father or Nemo's apprehensive father, Po's adoptive father, or Belle's eccentric father. In fact, it is only in the second half of the twentieth century that the father figure becomes progressively central in the care and growth of the child. As a result of cultural, social, and economic processes, fathers have tripled the time spent in direct play and care activities with their children from 1965 to today, and fatherhood is increasingly a key dimension of male identity. In the past decade, scientific research has also started to focus specifically on understanding the psychobiological dimensions that characterize motherhood, fatherhood, and their respective behavioral dimensions. This field of study has grown thanks to the study of biparental species and it also reflects the socioeconomic and cultural changes that have led to the coining of the term "involved dad" to identify a father who participates in all child care activities and not just indirect activities. The study of the psychobiology of parenting, that is, the processes of embodied parenthood, can also help understand how parents and their children can enter states of synchrony, even from the earliest moments of life.

Psychobiology of maternal caregiving

The neural and hormonal substrates of maternal behavior have been inherited and well-preserved through evolution in various mammalian species. Many of the neural circuits that underlie maternal behavior are organized into two broad systems that mediate responses of defense and approach towards offspring and sensory stimuli from the offspring. The transition to the acquisition of maternal behavior is

DOI: 10.4324/9781003479314-14

characterized by the suppression of circuits of avoidance, defense, or aggression towards such stimuli. Paternal caregiving and alloparental behavior, on the other hand, may have evolved independently in many mammalian species and may involve neural circuits only partially borrowed from the maternal brain.

In mammals, virgin females are generally averse to stimuli related to offspring (Numan, 2012), and pregnancy contributes to making females more sensitive and ready to assume and display caretaking behaviors in the postnatal period. Hormones present during pregnancy and childbirth promote the production and availability of maternal milk and facilitate the approach and acceptance of offspring for the new mother. Male mammals who do not experience the same hormonal changes related to pregnancy often avoid offspring and may even commit infanticide in certain environmental conditions. However, in species of mammals where a cooperative caregiving system has developed, such as in prairie voles and humans, aversion to stimuli related to offspring is not as widespread in females or males (Hrdy, 2016). The emergence of maternal caregiving behavior is characterized by positive evaluation and acceptance of stimuli related to the offspring and increased motivation for approach and care toward the progeny (Numan, 2012). In rodents, the olfactory bulb, amygdala, and anterior region of the hypothalamus are part of a circuit that contributes to aversive reactions toward offspring (Bridges, 2015). Pregnancy hormones contribute to turning off this circuit by inhibiting maternal avoidance response through the action of the medial preoptic area. At the same time, an approach and caregiving circuit include dopaminergic innervations of the limbic system by the medial preoptic area (Stolzenberg & Champagne, 2016). Along with the bed nucleus of the stria terminalis, this area receives stimuli related to the offspring and activates the ventral tegmental area, signaling to the nucleus accumbens the motivation for approach and caregiving through dopaminergic activation of the reward circuit. In fact, the activity of these brain areas is high in lactating female mice, while the activity of the bed nucleus of the stria terminalis is reduced (Matsushita et al., 2015). Oxytocin contributes to making stimuli from the offspring attractive to the mother by stimulating the auditory cortex of the left hemisphere (Marlin et al., 2015). Similarly, arginine vasopressin in the medial preoptic area increases the likelihood that the mother will engage in searching movements towards the offspring (Bayerl et al., 2016).

Research has shown that paternal behavior in mammals is highly variable both phenotypically and biologically, perhaps even more so than maternal behavior (Rogers & Bales, 2019). In fact, the brain adaptation to the transition to fatherhood in males seems to proceed in a less canalized way and under less hormonal control than in females, whose changes already originate during pregnancy. Conversely, in males, these neurobiological variations seem to be more susceptible to environmental influences, such as exposure to the partner's pregnancy and direct involvement in interacting with the offspring (Feldman et al., 2019; Grumi et al., 2021). This exposes paternal behavior to broader and less predictable influences from individual characteristics and the value and cultural systems of the group to which the individual belongs. In other words, the paternal phenotype can be considered more plastic than the maternal one.

Paternal caregiving is present in some fish, birds, and insects and only in 3–5% of mammals. Since parental investment is costly, the question is: why have some species developed this behavior? It tends to be observed when it improves offspring survival, particularly in monogamous species - where it is associated with higher probabilities of offspring survival, larger litter size, and faster offspring growth (Rogers & Bales, 2019). It is possible that once paternal caregiving has emerged, it stabilized the monogamous mating system and facilitated the emergence of complex socio-cognitive behaviors (Stockley & Hobson, 2016). It is also possible that prolonged exposure to offspring reduced the risk of infanticide by males or that males started defending access to the female specimen to ensure reproductive success in a particularly competitive context (Fernandez-Duque et al., 2009).

Is there a paternal brain?

As previously mentioned, it is believed that the behavioral phenotype of fatherhood is likely dependent on the same anatomical, functional, and neuroendocrine pathways observed in the context of maternal behavior. However, the presence of greater plasticity in the paternal phenotype leaves open the possibility that some observed differences in behavior may reflect different circuits and pathways in the central nervous system and hormonal influences. For example, a genome-wide study conducted between two different mouse species—*Peromyscus polionotus* and *Peromyscus maniculatus*—revealed that 12 genomic regions control the expression of the parental behavioral set, but 8 of these regions show differences between mothers and fathers, suggesting that maternal and paternal behavior may have followed different evolutionary paths in these rodent species (Bendesky et al., 2017). The study of the neural and hormonal correlates of paternal behavior has so far been mainly conducted on animal models of primates, such as the common marmoset and the tamarin, and on biparental rodents, such as the prairie vole, the mandarin vole, the California mouse, the dwarf hamster, and the Mongolian gerbil.

Some of the brain circuits and mechanisms involved in maternal behavior were anticipated in the previous paragraph. Even in the model of paternal behavior in rodents, the medial preoptic area of the hypothalamus appears to be involved in promoting the male's sensitization to offspring, i.e., the production of hormones and the expression of caretaking behaviors following prolonged exposure to the pups. Lesions of this area inhibit paternal caregiving behavior in the California mouse (Lee & Brown, 2002), and the activity of the medial preoptic area is high in individuals of the same species after prolonged exposure to offspring (De Jong et al., 2009). The amygdala, pallidum, and lateral septum are other regions involved in paternal behavior in rodents (Lee & Brown, 2007). In the prairie vole, exposure to offspring in adult males induces the production of a marker of neuronal activity in various brain regions, including the medial preoptic area, amygdala, and lateral septum (Kirkpatrick et al., 1994). These regions can be considered the subcortical area of the paternal brain (Feldman et al., 2019).

Studies conducted using the paradigm of separation from the father figure have shown in animal models that when the father is removed in a biparental species, negative consequences emerge for the neurological, behavioral, and social development of the offspring (Birnie et al., 2013; McGraw & Young, 2010; Yu et al., 2012). In the California mouse, paternal deprivation alters synaptic density and decreases the expression of proteins involved in neurogenesis (Bredy et al., 2007). In the medial prefrontal cortex of this mouse, paternal figure deprivation can induce changes in the modulation of the dopaminergic system and in the glutamate-driven excitation of pyramidal neurons. In the degu, the loss of the father figure reduces synaptic connectivity in the sensory cortex in males (Pinkernelle et al., 2009), indicating that sensory stimulation provided by contact with the father may promote the development of this brain area in rodents. Additional effects have been observed at the level of molecules that induce neural plasticity. For example, in the mandarin vole, paternal deprivation reduces the density of hippocampal neurons containing cortisol receptors in both males and females (Wu et al., 2014), and prolonged separation from the father is associated with reduced expression of dopaminergic receptors (Yu et al., 2012). In the prairie vole, the removal of the father induces long-lasting increases in BDNF expression and its receptors (Tabbaa et al., 2017). In the animal model, structural changes also occur, such as neurogenesis or alteration of the density of synaptic branches in the hippocampus, which are stimulated by olfactory stimuli from the offspring (Mak & Weiss, 2010) and direct caregiving experience (Kinsley & Lambert, 2008). In the California mouse, male involvement in direct caregiving activities, for example, facilitates increased dendritic branching in the dentate gyrus in the paternal brain, as well as in that of the offspring (Glasper et al., 2016).

In humans, functional resonance studies have highlighted numerous areas that seem to be responsive to visual or auditory sensory stimuli related to the child (Feldman et al., 2019). Brain activation and gray matter volume change in the same subcortical circuits previously highlighted in the animal model. However, other cortical areas appear to be involved in the paternal brain in humans. These are areas involved in circuits regulating emotions, mentalization, and sensorimotor simulation of observed goal-directed movement in others and therefore may be involved in promoting parenting behaviors such as understanding the child's behavioral state and intention, responding contingently to their emotional signals, and initiating interactive exchanges characterized by synchrony and attunement. The activation of these areas is further facilitated by hormones already associated with maternal behavior, such as oxytocin and vasopressin, as well as cortisol and testosterone (Mascaro et al., 2013; Nishitani et al., 2017). Two studies that specifically investigated differences in brain activation in response to their own child's images in mothers, fathers in heterosexual couples, and primary caregiving homosexual fathers without maternal involvement, showed similar activations for all three groups. However, mothers showed up to four times greater activation of the amygdala, a subcortical area, while fathers seemed to activate the superior temporal sulcus more, a cortical area with important implications for socio-cognitive functions such as mindreading and understanding intentions (Atzil et al., 2012). In another study, Ruth Feldman and colleagues (Abraham et al., 2014) investigated the different brain activations in response to images of their own child in mothers, fathers in heterosexual couples, and homosexual fathers who were primarily involved in caring for the child without maternal involvement. Similar activations emerged for all three groups. However, mothers showed up to four times greater activation of the amygdala compared to fathers, while fathers showed greater activation of the superior temporal sulcus. This activation was similar in fathers who were considered primary or secondary caregivers, while only in homosexual fathers, amygdala activity was similar to that observed in mothers. These results contribute to outlining a picture where, although the paternal brain may largely mirror the maternal brain, some regions appear to reflect different activation patterns. In particular, the role played by the cortex and the superior temporal sulcus suggests that, in fathers, the brain response may be more accurately defined as a top-down response, where observed behavior is not a direct consequence of evolutionarily inherited brain circuits shared with other mammals but depends on how brain plasticity has adapted to a cultural context where direct involvement in child care has grown.

Hormonal plasticity

There is a substantial scientific literature supporting the idea that maternal behavior is accompanied and prepared during pregnancy by changes in oxytocin levels (Feldman & Bakermans-Kranenburg, 2017). Oxytocin is a hormone produced by various nuclei located in the hypothalamus (Uvnas-Moberg et al., 2005) and has been associated with the early emergence of maternal sensitivity behaviors and the development of a sense of attachment to the fetus and newborn in women (Feldman & Bakermans-Kranenburg, 2017; Kendrick, 2000). The study of oxytocin's role in paternal behavior has received less attention, partly because men do not experience the same hormonal changes that accompany pregnancy. However, this difference may suggest that paternal behavior is more flexible and sensitive to hormonal variations induced by context and direct exposure in child care (Abraham et al., 2014).

In mammals, the oxytocinergic system has been associated with facilitating caring behaviors in both females and males (Ross & Young, 2009). In mandarin voles, fathers show an increase in oxytocin levels compared to males without offspring, and those who are more involved in caring for their offspring show even higher levels (Wang et al., 2015). Experimental manipulation of oxytocin levels can also induce similar behavioral effects. The administration of oxytocin seems to promote defense, caregiving, and physical proximity behaviors toward offspring in mongooses (Madden & Clutton-Brock, 2011) and marmosets (Saito & Nakamura, 2011; Woller et al., 2011). In biparental species such as meadow voles, fathers show elevated oxytocin levels in brain regions involved in emotional responses to the pups, such as the lateral amygdala (Parker et al., 2001). Male virgin grassland voles that have not been previously exposed to females exhibit an increase in plasma oxytocin levels when exposed to prolonged contact with offspring (Kenkel et al., 2014). These studies suggest that oxytocin is widely involved in the emergence of paternal behavior in mammals.

A recent review of the literature (Grumi et al., 2021) extends this evidence to humans. This study analyzed 24 studies conducted on more than 800 fathers and provided a still partial but intriguing picture of the role of oxytocin in paternal behavior in humans. Oxytocin concentrations are higher in fathers than in childless males (Feldman et al., 2012; Mascaro et al., 2013), and fathers show similar oxytocin levels compared to mothers (Gordon et al., 2010; Miura et al., 2015). Furthermore, oxytocin levels seem to increase only after the baby is born (Cohen-Bendahan et al., 2015), suggesting that exposure to direct contact with the child influences the regulation of the serotonergic system in fathers, although psychological adaptation to fatherhood may begin as early as pregnancy (Lindstedt et al., 2020). Some paternal behaviors, such as effective object presentation and engaging in physical play activities, are supported by elevated oxytocin levels (Apter-Levi et al., 2014; Feldman et al., 2010; Naber et al., 2010; Weisman et al., 2012). In risky settings, such as with preterm babies, early physical contact between father and child, such as skin-to-skin contact in neonatal intensive care, can increase oxytocin secretion in fathers (Cong et al., 2015). Fewer studies have focused on co-regulation or intergenerational transmission of oxytocin levels between fathers and children. The quality of paternal behavior experienced by fathers in their childhood was associated with oxytocin levels measured after an interactive exchange between fathers and their 4- and 6-month-old children (Feldman et al., 2012). In addition, higher levels of oxytocin synchrony between fathers and children have been reported in dyads that also showed elevated levels of behavioral attunement (Feldman et al., 2010).

Although research on the associations between parental behavior—maternal and paternal—and oxytocin has methodological and technical limitations (Grumi et al., 2021; McCullough et al., 2013), the study of the endocrine components of parenting seems to confirm the hypothesis that paternal behavior is more influenced by experience, such as direct involvement with the child. Future studies in this area will further help us understand how parenting—in both mothers and fathers—emerges from a complex interplay of endogenous neurobiological preparations programmed by evolution and social experiences shaped by cultural and social practices.

Synced dancers

The dyadic interaction between parent and child is characterized by a complex set of processes that contribute to defining an organization, an assembly grammar, that proceeds seemingly disorderly through continuous cycles of coupling or matching states, dyadic repairs, and new separations (Provenzi et al., 2015a). This process can be broken down into various types of dyadic encounters: matching, mismatching, repair, synchrony, contingency, coordination, tuning, and mirroring (Provenzi et al., 2018e). The emergence of reliable and moderately predictable mutual co-regulation is one of the implicit goals of early parent-child interaction and is also driven by the very early neurobiological and physiological regulations that characterize the dyad (Davis et al., 2018; Provenzi et al., 2019b). Ruth Feldman (2006) suggested that patterns of behavioral co-regulation are already negotiated at the physiological level; in other words, the observed rhythm after birth would depend on a biological, visceral dance.

From birth, the mother begins to engage in a series of species-specific behaviors that include eye contact exchanges, special vocalizations—like motherese—displays of specific emotional states, affectionate touch, and contingent responses to the child's signals. The child is biologically programmed to perceive and respond to maternal signals, initiating initial social exchanges marked by specific rhythms (Feldman, 2012). Maternal oxytocin levels in the first trimester of pregnancy have been associated with the quality of postnatal maternal behavior and the coordination of maternal responses to the child's communicative signals (Feldman et al., 2007). In a longitudinal study, Ruth Feldman (2012) measured the degree of behavioral synchrony in mother-child dyads at 3, 6, 12, and 24 months, at five and ten years. From birth, interactive synchrony was significantly associated with the child's socio-emotional development at the subsequent assessment points, and vice versa, suggesting a bidirectional relationship between interactive synchrony and the development of the child's emotional regulation skills. Furthermore, the behavioral synchrony observed at 3–4 months significantly predicted self-regulation abilities, empathic development, and symbolic competencies up to adolescence.

In the parent-child interaction, these psychobiological co-regulations can be observed at the level of various systems. From a neuroendocrine perspective, it has been hypothesized that the regulation of the hypothalamic-pituitary-adrenal (HPA) axis in parent and child is a protective and developmental factor for other interactive, social, and emotional functions (van Bakel & Riksen-Walraven, 2008). Bernard and colleagues (2017) recently proposed considering the early tuning of the HPA axis between parent and child as a continuous process reflecting the mutual and integrated efforts of the two interactive partners in regulating socio-emotional stress situations. Hibel and colleagues (2015) observed mother-child dyads during a series of specially designed tasks to stimulate frustration, anger, and stress responses in the child at 7, 15, and 24 months. The tasks were progressively more stressful, starting with a barrier task where an object of interest to the child is placed behind a transparent barrier that is difficult to reach and ending with a brief episode of physical restraint of the child's hands. Salivary cortisol samples were collected before and after the observational procedure. At all three ages, mothers and children showed a significant correlation between their respective levels of salivary cortisol; however, the association became more evident as the level of stress elicited by the observational procedure increased.

In another study, Provenzi and colleagues (2019b) quantified the co-regulation of salivary cortisol levels between mother and child before and after the still-face procedure in dyads of full-term and preterm-born children. Salivary cortisol was collected at four time points: before the procedure (baseline), as well as 10 (initial reactivity), 20 (late reactivity), and 30 (recovery) min after the end of the reunion phase. Co-regulation was measured in two different ways: the correlation of salivary cortisol levels at each individual time point and the synchrony in cortisol level variations across the different time points. Considering the correlation of salivary cortisol levels at each of the four time points, a significant association emerged only for the full-term group before the still-face procedure and after 20 and 30 min. However, in the preterm-born children dyads, there were no significant differences in the trajectories of mother and child between baseline and recovery, while in full-term-born children dyads, the values tended to diverge between baseline and initial reactivity, as well as between initial reactivity and late reactivity. The results suggest that adrenocortical co-regulation in the mother-child dyad may be particularly complex and may reveal different mechanisms of tuning the biological rhythms of the two interactive partners depending on the method of measurement. In this case, using two different indices, it is possible to observe that in full-term-born children dyads, the mother and child are actively involved in regulating each other's HPA axis in a complementary way, as if their respective neuroendocrine systems were part of a dynamic dyadic system: the child increases cortisol levels, while the mother decreases them. Conversely, in preterm-born children dyads, there is maximization of correlation at a given moment, and the two interactive partners seem to show the same trajectory of response to the still-face, with a mimicking decrease in salivary cortisol levels. Although it is not possible to determine whether one or the other pattern of adrenocortical co-regulation is more protective for the child's development, it is evident that early experiences, such as neonatal intensive care hospitalization, can significantly alter the biological synchrony processes between mother and child.

The co-regulation of the parasympathetic nervous system has also been studied, although less extensively, in the mother-child dyad. In the first two studies that investigated parasympathetic co-regulation in mother-child dyads (Ham and Tronick, 2006; Moore et al., 2009), a pattern of inversion was observed in the response of vagal tone to the still-face in mothers and children: while children showed on average a suppression of vagal tone during the still-face phase and a return to baseline levels during the reunion, mothers showed the opposite pattern. An initial suppression of RSA in mothers between the habituation phase and the start of the play phase suggested that during the still-face phase, mothers were not engaged in interaction; however, this finding is surprising considering that witnessing one's own child expressing frustration can be equally stressful for the mother (Tronick, 2003). In a subsequent study, Moore and colleagues (2009) explored parasympathetic co-regulation in 6-month-old mother-child dyads using the maternal still-face procedure. In this study too, mother and child exhibited opposite regulation patterns, with a suppression of vagal tone followed by a return to baseline values in the child and an increase in RSA followed by the same recovery in mothers. Physiological co-regulation can also be observed using thermal cameras. The Italian group led by Tiziana Aureli (Ebisch et al., 2012) used this tool to observe the autonomic response in 12 mother-child dyads aged between 38 and 42 months. The observational procedure was the so-called mishap paradigm, where the child is asked to play with a toy that has been manipulated to break when the child starts using it. In this study, a sympathetic system response—a sudden and marked reduction in body temperature accompanied by sweating—was observed in the child after the toy broke. This response was partially recovered when the child was comforted by the adult. The same response was observed in the adult, and activations in specific areas of the face, the tip of the nose and the jaw region, were correlated in mothers and their children. These results suggest again how dyadic processes of physiological co-regulation, like those of an adrenocortical nature, can reveal how the dyad functions as a single dynamic system, where each participant's parameters are communicated synergistically.

Adrenocortical co-regulation has also been studied—although less extensively—in the father-child dyad. Mills-Koonce and colleagues (2011) measured the association between behavioral measures of paternal sensitivity and HPA axis responses to stressful conditions at 7 and 24 months. Lower paternal sensitivity was significantly and positively associated with higher salivary cortisol levels 20 min after stress. More recently, Saxbe and colleagues (2017) reported a significant correlation between paternal and child salivary cortisol levels in preschool children during a series of socio-emotional stress tasks. In a previous study, a correlation in basal cortisol levels emerged for the father-child dyad, but only in the evening hours (Stenius et al., 2008). More recently, Brenda Volling's group (Bader et al., 2021) studied the co-regulation of salivary cortisol secretions in father-child dyads during the strange situation procedure in 12-month-old children. Salivary cortisol was collected from both before (T1) and after 20 (T2) and 40 (T3) min from the end of the procedure. Correlations conducted at each of the three moments were significant; however, evidence of time synchronization, between the moments, of cortisol levels was limited.

Interbrain wires

More recently, the availability of neuroimaging devices that allow the measurement of brain activity in vivo during the interaction between two subjects—hyper-scanning studies—has initiated a series of studies that are contributing to understanding how processes of central nervous system co-regulation can also emerge in the mother-child dyad. Using electroencephalographic (EEG) measures, Wass and colleagues (2018) assessed the presence of brain synchrony in mother-child dyads by alternating conditions of solitary and interactive play. During solitary play phases, the child's theta activity increased before shifting visual focus to another object. During interactive play moments, the parent's theta activity increased following the child's eye movements, indicating that the mother's brain activity is capable of supporting parental responsiveness to the child's exploration signals. The group of Victoria Leong (Santamaria et al., 2020) demonstrated that the synchronization of EEG signals between mother and child was stronger in the alpha rhythm when the mother expressed positive emotional states during the interaction, compared to when she displayed negative emotions. In a study conducted by Gianluca Esposito's group (Azhari et al., 2019), 31 mother-child dyads participated in an experiment where both were exposed to videos that elicited positive or negative emotions. The videos were short 1-min clips from cartoons such as Brave, Peppa Pig, and The Incredibles. The children were about three years old and seated on their mother's laps while both watched the screen. Brain activity in mothers and children was measured using functional near-infrared spectroscopy (fNIRS). In dyads with mothers reporting higher levels of parental stress, less dyadic brain synchrony was observed in the prefrontal cortex, a region extensively involved in social cognition tasks and interpreting others' mental states. Thus, maternal mental state and parental stress levels may also influence the quality of neurophysiological co-regulation in the dyad, particularly in areas involved in socio-cognitive and relational functioning. Recently, the same group (Azhari et al., 2020) replicated these results using father-child dyads exposed to the same experimental paradigm. In another study (Krzeczkowski et al., 2020), a synchronized system with two EEG headsets was used to assess dyadic brain synchrony in typically developing 9-month-old children and their mothers during an emotional task. The children were seated in a highchair facing the mother. After a 5-min resting state phase, the dyads were exposed to two conditions in which musical stimuli—the Second Movement of Vivaldi's Spring and an excerpt from Disney's Peter and the Wolf—were played to elicit specific emotional states. Frontal EEG signal asymmetry was evaluated in both conditions, as it has been shown to be a sensitive index of emotional valence stimuli (Coan & Allen, 2004; Diaz & Bell, 2012). The mothers' tendency to display proximity-seeking or avoidance behaviors was assessed with questionnaires measuring stable personality traits. In dyads with mothers characterized by high social avoidance, maternal frontal asymmetry during resting state predicted that of the child during exposure to both emotional stimuli. Therefore, co-regulation and reciprocal influence between maternal and infant brain states may be facilitated under specific conditions characterized by a maternal avoidant style. However, the study of brain-to-brain tuning processes between mother and child is still in its infancy, and further research is needed to better understand the mechanisms and environmental conditions that promote or inhibit brain-to-brain communication in the mother-child dyad. Maria Gartstein recently showed how it is possible to study brain co-regulation in mother-child dyads using the still-face procedure as well (Perone et al., 2020). In this study, 10 mother-child dyads participated in the still-face paradigm when the children were 10 months old, and frontal alpha rhythm asymmetry was measured using EEG. The dyads showed a right lateralization of frontal asymmetry during the procedure. In addition, in dyads with more sensitive mothers, frontal asymmetry was more pronounced, which may reflect a greater presence of dyadic tuning or involvement. Although Perone et al.'s study (2020) is solely descriptive, studying brain co-regulation processes using well-validated dyadic interaction paradigms, such as the maternal still-face procedure, seems to be a promising avenue to advance our understanding of early neurobiological synchrony processes in the parent-child dyad.

12 Memories of trauma

Mary Ellen Wilson was born in 1864 in the Hell's Kitchen neighborhood of New York City to a couple of Irish immigrants. After her father, Thomas, was killed in the Civil War, her financially struggling mother entrusted the child to the Department of Charities. Soon after, a couple named Mary and Thomas McCormack claimed to be Mary Ellen's biological parents, and they obtained legal custody of the child. Mr. McCormack died shortly thereafter, and Mary McCormack married Francis Connolly. One of the black and white photographs of Mary Ellen Wilson available online depicts her with a large gash on her right cheek, and her arms and legs show obvious signs of beatings and abuse. The photograph was taken when Mary Ellen was 10 years old. Mary Ellen had lived in servitude, taking on responsibilities well beyond her age. She had never played with other children, hardly ever left the house, and was often confined to her locked room without a bed or toys. Neighbors later reported that they had never seen her but often heard her crying.

In 1874, despite the United States being a democratic country with a Constitution and a Bill of Rights, institutions were largely powerless to defend the rights of a child. Children were considered little more than property, and while there were laws protecting them, they were rarely enforced, and parents were mostly immune from criminal charges. Mary Ellen's case was reported to a Methodist missionary named Etta Angell Wheeler, who regularly visited Hell's Kitchen. When Miss Wheeler saw the condition of the little girl, she was shocked. At nine years old, Mary Ellen looked like a five-year-old, and despite it being winter, she was dressed in light clothes without socks or shoes while washing dishes on a wooden box. Miss Wheeler tried to contact the police, but without direct testimony, it seemed that nothing could be done, until she was introduced to Henry Bergh.

Mr. Bergh was the son of an American shipbuilder and diplomat in Russia and had inherited great negotiating skills. He had founded the first American society for the prevention of animal cruelty in 1866. Thanks to this organization, he had repeatedly arrested captains of commercial ships for mistreatment of animals, invented the ambulance for injured horses, and created clay pigeons for hunting practice, saving hundreds of birds from unnecessary death. He had also sued—and won—against the Barnum circus for the mistreatment of exotic animals. When Miss Wheeler contacted him, he responded cautiously but decisively. He wrote to a lawyer named Elbridge Gerry and managed to convince Judge Lawrence of the

DOI: 10.4324/9781003479314-15

New York Supreme Court to take up the case within 48 hours. When the police officers took custody of Mary Ellen, she was so frightened of the outside world that she used the lollipop given to her by one of the officers as a weapon to defend herself. Bergh and Gerry founded the first New York Society for the Prevention of Cruelty to Children in 1874. In its first year of activity, this organization investigated more than 300 cases of suspected abuse and mistreatment. At the age of 24, Mary Ellen Wilson married a widower with whom she had two biological daughters and adopted a third. Her biological daughters both became teachers, and her adopted daughter became an entrepreneur. Her grandchildren described Mary Ellen as a loving, kind, and fairly permissive grandmother. She passed away at the age of 92 in 1956.

Today, child maltreatment, including physical or sexual abuse, psychological violence, neglect, and exposure to domestic violence, is still a widespread problem worldwide. Prevalence estimates vary considerably based on different definitions of maltreatment, data collection methods, sources, and the level of disclosure. Many cases of childhood maltreatment remain unknown and are often discovered very late. This makes it difficult to systematically compare data obtained in different countries. The estimated prevalence of sexual abuse varies between 0.4% (from third-party sources or registries) and 12.7% (when reported by the subjects themselves). In Italy, out of 1000 minors taken care of by the services, 20% have a history of mistreatment. Out of a sample of more than 90,000 minors, 47% reported physical or emotional neglect, 19% experienced violence, 14% reported psychological maltreatment, and a percentage between 9% and 4% experienced other forms of physical abuse, inappropriate care, or sexual abuse.

The application of the lens of developmental psychobiology to the study of child maltreatment is helping us understand how early traumatic experiences can have long-lasting effects on the child's behavior, psychology, emotions, and cognition. Changes in psychoneuroendocrinology, affective neuroscience, and behavioral epigenetics are helping to identify the markers that can influence how our physiology and biology learn from trauma and prepare us for future stressful events, for better or worse. However, many traumatic events of childhood are only recognized and addressed later, often years later, which has implications for psychobiological research on childhood maltreatment. The data collected are often self-reported by the participants, and the surveys are retrospective, making the evidence correlational. Other traumatic events, such as natural disasters or potentially traumatic events that transcend an individual's life, can serve as quasi-experimental study contexts, allowing researchers to investigate the biological impact of trauma and its contribution to developmental programming more prospectively and with a relative degree of control over some confounding variables. Therefore, in this chapter, after presenting a brief and non-exhaustive review of psychobiological studies conducted in the context of child maltreatment, studies related to four different "large-scale" traumatic experiences will be presented: the terrorist attack on the World Trade Center in 2001, the ice storm in Canada in 1998, the Dutch famine of the forties of the twentieth century, and the intergenerational effects observed in the children or grandchildren of Holocaust survivors. Studying the medium and long-term consequences—even, precisely, intergenerational consequences—of these events can be of great help in understanding how adverse experiences affect our biology and identifying potential biomarkers of stress and future preventive and therapeutic interventions.

Psychobiological scars

It is possible to observe the biological and physiological consequences of childhood trauma through various systems involved in stress regulation. The Trier Social Stress Test (TSST) is a widely used procedure to observe and measure behavioral and neuroendocrine stress response in laboratory settings with adult subjects (Heim et al., 2000). Women with a history of abuse, even in the absence of a major depression diagnosis, show a significant increase in the production of adrenocorticotropic hormone (ACTH), cortisol, and heart rate in response to the TSST compared to controls (ibidem). Depressed women without a history of abuse do not display the same pattern of stress hyper-reactivity in the laboratory. Other studies have replicated these results (Heim et al., 2002) and elucidated the underlying mechanisms in animal models (Rao et al., 2008), although some researchers have reported an opposite effect of hypothalamic-pituitary-adrenal (HPA) axis hypo-responsiveness (Carpenter et al., 2011).

Disinhibition of central corticotropin-releasing hormone (CRH) and noradrenaline release might be one of the mechanisms involved in the emergence of hyper-reactivity. This inhibition could be the result of changes that occur in the feedback control mechanisms at the glucocorticoid receptors (GR) level in the HPA axis. The combined dexamethasone and CRH administration test was developed to test the HPA axis's ability to maintain efficient feedback control even under stress conditions (Heuser et al., 1994). In this test, a high dose of dexamethasone is orally administered in the evening, resulting in HPA axis shutdown. The next day, CRH is injected to simulate a stressful condition. In healthy subjects, HPA axis activity remains relatively subdued, reflecting adequate action and sensitivity of hippocampal GR receptors. In depressed subjects, however, inhibition of the feedback control effect and neuroendocrine hyperactivity, quantified by glucocorticoid production, are observed. This effect is even more pronounced in subjects with a history of abuse or maltreatment (Heim et al., 2008). This glucocorticoid resistance by GR receptors could, in turn, enhance the action and activity of pro-inflammatory pathways, such as excessive cytokine production in response to stress (Bierhaus et al., 2003; Danese et al., 2008).

Several regions are expected to show deficits or alterations in memory in maltreated children (Howe et al., 2006). Indeed, alterations in HPA axis functioning and regulation can influence cognitive and mnemonic performance (Cicchetti et al., 2010), and maltreated children who develop posttraumatic symptoms may also present with partial memories (Howe et al., 2006). Dante Cicchetti is one of the leading researchers in memory processes within the context of childhood maltreatment. In 2004, he examined performance in a false memory recognition task in maltreated children and controls without a history of maltreatment (Howe et al., 2004), documenting the absence of significant differences between the two groups. The results are consistent with other similar studies conducted on subjects with a history of sexual abuse or post-traumatic stress disorder (PTSD) and childhood maltreatment (Beers & DeBellis, 2002). In 2010, Cicchetti further replicated these results, but by investigating different types of maltreatment in association with alterations in HPA axis regulation in basal conditions, he documented some individual differences. Children who had experienced neglect and/or emotional maltreatment and showed low morning cortisol levels had a higher risk of false memory, suggesting that the effects of neuroendocrine system regulation on the cognitive development of children exposed to early trauma may also depend on the nature of the trauma and how it is processed by the child (Valentino et al., 2008).

Consequences of early trauma have also been observed at the central nervous system level. It is important to note that the human brain is highly plastic in the early stages of life. Different brain regions may present varying degrees and rates of maturation (Lupien et al., 2009). The hippocampus reaches maturity at around five years of age, while the amygdala matures much earlier, around two to three years. At the same time, the prefrontal cortex continues to mature until adolescence and adulthood, and functional connections between these different regions continue to grow and change based on experience throughout childhood and adolescence (Teicher et al., 2016). The hippocampus plays a critical inhibitory role in HPA axis activity and is involved in the conditioning of the response to fear and environmental threats. Prolonged states of stress can result in decreased excitability of hippocampal neurons, dendritic atrophy, and apoptosis, and low hippocampal volume has been associated with major depression (Lupien et al., 2009), as well as in subjects with a high number of traumatic events in childhood (Teicher et al., 2016; Vythilingam et al., 2002). The prefrontal cortex mediates executive functions and motor activation directed towards a goal and is also involved in emotional regulation. Reduced prefrontal cortex volume is a characteristic found in PTSD (Ressler & Mayberg, 2007) and subjects with early histories of maltreatment (Tomoda et al., 2009). Functional neuroimaging studies have shown that early traumatic events can be associated with amygdala hyperactivity in response to threatening stimuli (Dannlowski et al., 2012; Grant et al., 2014; Tottenham et al., 2011). Reduced connectivity between the amygdala and the prefrontal cortex has been documented in subjects with a history of repeated traumas (Govindan et al., 2010). Heim and colleagues (2013) have also documented alterations in the somatosensory cortex that seem to be specific to the type of trauma experienced. For example, pronounced cortical thinning of the somatic area relative to the genitals was observed in subjects with sexual abuse, and thinning of the anterior cingulate cortex was associated with emotional abuse. Similar results have been documented for the visual and auditory cortex in subjects exposed to domestic violence (Tomoda et al., 2012).

Pariante's group (Danese et al., 2007) studied the effects of maltreatment on the immune system in a longitudinal cohort—the Dunedin Study—initiated in New Zealand in 1972, with 1000 participants followed longitudinally until adulthood. The results of interest for this chapter concern the measurements performed when subjects were 32 years old. Exposure to maltreatment was significantly and directly associated with higher levels of CRP in adulthood; participants with a likely history of maltreatment showed only a slight increase in CRP levels, which were markedly elevated in subjects with a documented history of severe and repeated maltreatment. Since it is possible that individuals exposed to early traumatic events may also display psychiatric consequences, such as depressive symptoms or psycho-pathological conditions, Danese and colleagues (2008) also investigated the joint effect of early traumas and depressive symptoms in adulthood. In their study—also conducted through the Dunedin Study longitudinal cohort—they divided the sample into four groups: subjects with a history of maltreatment but without a depression diagnosis; subjects with a history of maltreatment and depressive symptoms; depressed subjects without a history of maltreatment; subjects without childhood traumatic events or depressive symptoms. CRP levels were elevated in 18% of the latter group—the control group—25% of the group with depression but without a history of maltreatment, 30% of those with maltreatment but without depressive symptoms, and 37% of those with a double-risk condition. The authors suggest that these results could help identify depressed patients with a traumatic history who might be particularly resistant to exclusive pharmacological therapy (Danese et al., 2011; Nanni et al., 2012).

Studies investigating the consequences of trauma at the level of specific genes have mostly focused on NR3C1, which encodes for cortisol receptors (GR receptors), on SLC6A4, which encodes for the serotonin transporter, and on FKBP5, a gene involved in the immune response and inflammation regulation processes. Seventy-four percent of the studies conducted on the NR3C1 gene have documented the presence of a hypermethylation profile in subjects with a history of physical, psychological, sexual abuse, or neglect (Bustamante et al., 2016; Cicchetti & Handley, 2017; Parade et al., 2016; Radtke et al., 2015). The association seems to be stronger for the earlier and more repetitive the maltreatment experiences (Cicchetti & Handley, 2017). In addition, maltreated individuals showing high levels of NR3C1 gene methylation tend to respond to psychological stress—TSST paradigm—with increased cortisol production (Alexander et al., 2018) and are also at higher risk of developing internalizing psychopathology (Peng et al., 2018). For the SLC6A4 gene, a higher methylation has also been extensively reported in subjects with a history of maltreatment (Booij et al., 2013; Vijayendran et al., 2012). On the contrary, a series of studies have documented low levels of FKBP5 gene methylation in maltreated subjects (Parade et al., 2017; Tozzi et al., 2018; Tyrka et al., 2015). Adults with a history of maltreatment have also shown shorter telomere length compared to control subjects, an effect that seems to be at least partially moderated by the number and chronicity of traumatic events (O'Donovan et al., 2011).

9/11

Although research on prenatal stress and the impact of postnatal adverse events suggests that the most significant consequences may generally be attributed to prolonged or repeated exposures, it is possible that even specific, particularly traumatic, unexpected, and painful events may result in lasting and significant psychobiological alterations. This is the case, for example, with what we are beginning to understand about the psychobiological consequences of the terrorist attack on the World Trade Center. The destruction of the Twin Towers on the morning of September 11, 2001, caused nearly 3000 deaths and was a traumatic event of global psychological implications (Palmieri et al., 2007).

Lederman et al. (2004) studied the effects of exposure to stress and the dust raised by the attack on the World Trade Center in 300 non-smoking women who lived within two miles of the site and were pregnant at the time of the terrorist attack. The women were recruited for the study between December 2001 and June 2002 from three different hospitals in the attack area. The time spent in the same geographic area was reported by the women for each week, and additional biological and physiological data related to maternal health and pregnancy were collected from medical records. The children of these women showed lower birth weight and length compared to children born to mothers living in areas farther from the attack site. Furthermore, women who were in the first trimester of pregnancy at the time of the terrorist attack gave birth to children with a lower gestational age and smaller head circumference compared to women whose pregnancies were at a more advanced stage. In another study, a sample of individuals living within four blocks of the World Trade Center was prospectively evaluated at seven and eighteen months after the terrorist attack (Dekel et al., 2013). Salivary cortisol samples were collected after asking participants to recall the events of the morning of September 11. Subjects with high levels of PTSD produced higher levels of cortisol in response to the memory of the traumatic event. Moreover, in males, neuroendocrine response and severity of PTSD symptoms were significantly correlated.

Rachel Yehuda investigated the relationship between maternal PTSD and maternal and infant salivary cortisol in dyads of mothers directly exposed to the September 11 terrorist attack (Yehuda et al., 2005). Salivary cortisol was collected from mothers and children upon waking up and just before bedtime. Lower cortisol levels were found in mothers who showed more severe PTSD symptoms and in their children. The effect was more evident in mothers who had been exposed to the attack during the third trimester of pregnancy. More recently, variations in DNA methylation levels have been observed in large genetic regions between blood samples obtained from women exposed to the trauma of the September 11 terrorist attack and blood samples collected from a control group before 2001 (Arslan et al., 2020). In another study, subjects with high PTSD symptoms showed higher methylation of genes involved in synaptic plasticity, regulation of oxytocin, and inflammatory processes (Kuan et al., 2017).

Below zero

In 1998, a series of ice storms hit a swath of land in Eastern Ontario, Quebec, and Nova Scotia, Canada. Dubbed the "Great Ice Storm," it caused severe damage to trees and power infrastructure, resulting in widespread and prolonged power outages. Ice accumulation on power lines led to the collapse of over a thousand electrical towers. Millions of people were left in black-out for days, and some even for several weeks. More than 400 emergency shelters were quickly set up, accommodating around 17,000 people. This climatic event was dramatic, causing 34 deaths, with more than 20 attributed to hypothermia, and severely impacting major urban centers like Montreal and Ottawa. The maple syrup industry, one of Canada's key productions, was nearly devastated. Insurance claims amounted to nearly a billion dollars. Businesses reported over 3 billion dollars in lost earnings, and close to 2 billion dollars were spent on hydroelectric infrastructure repairs. Over 40,000 people lost their jobs.

In the following months, Suzanne King and David Laplante launched the longitudinal Project Ice Storm to determine the nature and duration of the stress effects related to this dramatic event, particularly in children born to mothers who were pregnant in January 1998 or became pregnant within three months after the Great Ice Storm. Around 17% of these women reported high levels of stress two years after the event. Compared to women exposed during the third trimester, those in the first or second trimester of pregnancy gave birth to smaller and less gestationally mature babies, and maternal prenatal stress significantly correlated with birth head circumference. Children born to mothers exposed to the ice storm during the first trimester of pregnancy and whose mothers reported higher stress levels showed lower scores on the Bayley scales, a standardized tool for assessing early cognitive development. More recently, objective measures of difficulty experienced by the mother during pregnancy, such as job loss, days without electricity, home or self-damage, and time spent in a shelter, were found to be associated with epigenetic modifications observed in the child at 13 years old: altered DNA methylation levels in over 1500 CpG sites in genetic regions related to immune system functionality.

The hunger winter

The term "Hongerwinter" refers to a famine that occurred between 1944 and 1945 in the Netherlands as a consequence of the Nazi-imposed embargo actions. This decision came after the Allied forces invaded and occupied Antwerp in the autumn of 1944, and the Dutch government initiated a railway personnel strike to support the offensive against the Nazi occupiers. The famine lasted for about 7 months, from November 1944 to May 1945, and the official daily rations, recorded in historical records, remained stable each week for each individual. These rations were less than 1000 cal. The famine had a profound impact on the overall health of the population. In the city of Amsterdam, the mortality rate in 1945 doubled compared to 1939, and much of the cause was attributed to malnutrition resulting from the Nazi embargo (Banning, 1946). The Nazi army records remained intact, enabling the understanding of how a restricted intake of calories—proteins, fats, and carbohydrates—could be a cross-generational environmental risk factor.

Adults conceived during the Dutch famine showed an almost doubled risk of developing schizophrenia and personality disorders (Neugebauer et al., 1999; Susser et al., 1996). In addition, individuals exposed to the restrictions of the Hongerwinter had higher blood glucose levels and lower insulin secretions (de Rooij et al.,

2006; Lumey et al., 2009), elevated blood pressure response to stress, and a higher risk of developing heart diseases in adulthood (Roseboom et al., 2000). Subjects exposed in utero to the environment of the Dutch famine during the early weeks of gestation showed higher indices of brain aging (Franke et al., 2018). More than 60 years later, reduced levels of DNA methylation were observed for the IGF2 gene (Heijmans et al., 2008) and other imprinted genes (Tobi et al., 2014). Studies on the epigenetic and clinical consequences of the Dutch famine suggest that the long-term effects of a less than optimal early environment may contribute to linking adverse events exposure in one generation with the risk of developing diseases in the subsequent generation and that these effects are at least partly related to the timing of exposure during pregnancy (Heijmans et al., 2009).

Inheriting a nightmare

Much of what we know about the intergenerational transmission of trauma is due to studies that have focused on understanding the behavioral and clinical issues in the children of Holocaust survivors (Rakoff et al., 1966). As Rakoff wrote, even though the parents may not show evident signs of psychiatric distress, their children—all born after the Holocaust—exhibit significant signs of psychological distress, to the extent that "one would be led to believe that it was they—rather than their parents—who had suffered the torments of hell." Since then, hundreds of articles have appeared and have begun to show how the second generation of Holocaust survivors displayed the stigma of trauma: feelings of hyper-identification with their parents, low self-esteem, a tendency to interpret life events as catastrophic, concern for the recursiveness of trauma, behavioral and psychological disorders, and a constant state of hyper-vigilance. These are mostly the same symptoms that contribute to defining a diagnosis of PTSD and have been described, for example, in Vietnam War veterans (Ancharoff et al., 1998).

Studies in the field of psychobiology have revealed that the Holocaust survivors themselves may have reduced basal cortisol levels and an increase in GR receptor sensitivity (Yehuda et al., 1995). Even in the absence of traumatic experiences in childhood, the children of Holocaust survivors may show dysregulations of the HPA axis similar to those of their parents (Yehuda et al., 2002; Yehuda et al., 2007, 2000). Moreover, the presence of PTSD in mothers and fathers may have different consequences in the second generation. In fact, while PTSD symptoms in female Holocaust survivors are associated with lower cortisol levels in their children, the same association did not emerge regarding paternal psychiatric symptoms (Lehrner et al., 2014; Yehuda et al., 2007).

However, it is possible that the observed biological effects in the second generation may reflect their own experiences or exposures to stressful events; indeed, the children of Holocaust survivors appear to have a higher rate of childhood traumas compared to age-matched controls (Yehuda et al., 2001). The picture is further enriched by more recent epigenetic investigations. In the absence of maternal posttraumatic disorder, the children of male concentration camp survivors who developed PTSD show higher levels of methylation of NR3C1 (Yehuda et al., 2014).

In the same study, the children of male and female Holocaust survivors showed lower levels of methylation in the same gene. Additionally, both survivors and their children show altered methylation levels in the same region of the FKBP5 gene, irrespective of the presence of posttraumatic symptoms. The methylation levels in this region are significantly correlated between the two generations, but they appear to go in opposite directions: high methylation levels in parents, low in children (Klengel and Binder, 2015; Yehuda et al., 2016).

Concluding remarks

Studying the psychobiological consequences of early trauma—whether it is chronic and repetitive or punctual—allows us to understand how our life experiences can produce lasting memories in our bodies. Moreover, these researches also contribute to recognizing the toxic consequences of collective dramas, as in the case of ethnic minorities (Evans-Campbell, 2008; Pihama et al., 2014) or populations that have experienced the trauma of war (Azarian-Ceccato, 2010; Barron and Abdallah, 2015; Roth et al., 2014; Svob et al., 2016). Our body does not forget. This knowledge helps develop predictive models of the effects of trauma on individual developmental trajectories. Also, it highlights intergenerational consequences and can provide evidence of the mechanisms through which cross-generational transmission of psychopathology can occur. In the coming years, these studies may help identify individuals who may be at higher risk of developing a psychiatric disorder as a result of direct or indirect exposure to traumatic events, facilitating the initiation of timely and individualized preventive interventions.

Epilogue
Framing developmental psychobiology in complexity

Complex

One of the guiding principles of this book is the concept of complexity. The complexity of dynamic systems, a way to say that we can approach the psychobiological processes of the dyad from many different points of observation. We need to be aware that no matter what lenses we wear to observe human development—typical or at risk—we will never fully grasp its complex organization. The study of developmental psychobiology is somewhat like a short blanket: as we uncover new mechanisms at the interface between the organism and the environment, the spotlight of our observations inevitably casts shadows on other areas of the landscape. Thus, the questions grow exponentially as we gain partial answers. Complexity has significant implications for the clinical implications of developmental psychobiology. Anyone working with children and parents quickly learns that there is no book or manual that can make us feel competent or complete. Every time, we start from scratch. What should a young psychologist who begins working with parents and observes minutes of parent-child interaction learn? There are no grids or easy solutions; the best way to approach complexity is with simplicity. Simple is indeed different from easy, just as complex is different from difficult. Observing what happens is not easy, but it simplifies complexity. Only then can we seek a motive, a reason, or an explanation. Embracing complexity, reflecting on it when approaching a patient, whether a child or an adult, means approaching with the taste of simplicity while taking seriously the person in front of us.

Wired

A second guiding thread is the concept of connection. We are connections: not only neural, neuroendocrine, genetic, epigenetic connections, but also social, cognitive and affective wires. These connections are intertwined, and their discontinuity can only be identified for didactic purposes, as I have tried to do in this book. The attempt to understand how inter-human connections contribute to the complex development of neurobiological connections (and vice versa) is one of the most significant driving forces behind the curiosity for this science, developmental psychobiology, which I am deeply passionate about. This curiosity also resonates with

Louis Sander's question: how is it possible that we are separate as individuals and at the same time capable of being together in an interaction? It is precisely thanks to the implicit capacity for connection that characterizes us as living beings that we can study the biological traces of our movement in the world and our inevitable encounter with the physical and human environment that surrounds us. Similarly, it is in these very connections and biological imprints that we can glimpse possible insights and mechanisms of action for prevention and treatment interventions for those who present conditions of risk or developmental pathologies. If psychopathology develops within human connections, it is not absurd to identify these same human connections as the preferred place for therapeutic action. In 1999, Kandel argued that to the extent that psychotherapy and psychological consultation are effective and produce long-lasting changes in behavior, it is conceivable that this occurs through the process of learning that modifies gene expression by acting on the efficacy of synaptic connections and rewriting the anatomical pathways of interconnections between neurons in the brain. In other words, it is possible to hypothesize that psychotherapy is effective when it leads the components of a system to oscillate and modify their reciprocal positions toward levels that exceed the stability parameters of a pathological life solution, moving toward a new configuration of states, patterns, and constraints that are more adaptive.

Quasi-adapted

A third key concept is that of adaptation. It appears to be a clear word whose meaning seems easily shared by everyone. However, it is a word that one should be cautious about. What does adaptation truly mean? "Quasi-adapted" (*it. Quasi adatti*) is almost an apt title for a 2001 song by the Italian rock band "Tre Allegri Ragazzi Morti." It is a piece that resonates with those who, like me, experienced their adolescence between the late nineties and the early years of the new millennium. I appreciate the concept of "quasi-adaptation" because it reminds me that the process by which an individual—an organism—attempts to adapt in its living environment is never concluded or given once and for all. It involves a continual movement of how an organism comes to terms with its living environment, both phylogenetically and ontogenetically. Evolution takes chances and places bets on adaptations that may confer advantages for the species in an average, predictable environmental context. However, as technical and scientific progress accelerates, the human environment becomes less and less predictable and increasingly different from one generation to the next, potentially reducing our biology's ability to make accurate predictions or investments that are not in the dark. At the same time, during the first years of life, the child interacts with an environment that contributes—thanks to the neuroplasticity of different biological systems—to define developmental trajectories, maximizing complexity, and ensuring the best possible level of coherence of the self. It becomes evident that the word "adaptation" begins to sound strange when applied to such a context. Perhaps a more effective and less semantically dangerous way of referring to this process is through "biological learning." This learning takes place outside of our awareness, being implicit in nature. A classic example of implicit learning involves learning to ride a bike. We learn implicitly by doing it—falling, sometimes—and approximating the complex set of movements, balances, and postural controls as we practice. There is no instruction booklet for riding a bike; it happens through implicit learning. Similarly, biological learnings are very much the same, but they are clearly inscribed in modifications of neuroplasticity or epigenetic plasticity. They allow us to learn to survive and—even better—to thrive within the context of life in which we find ourselves. From this perspective, paraphrasing Watzlawick, for our biology, we cannot fail to learn. Therefore, it is theoretically possible to describe the evolutionary trajectory of an individual through the modifications induced by their biological learnings at the neuroendocrine, physiological, epigenetic, and behavioral levels. An example of this type of biological learning is clearly demonstrated in social memory studies for the still-face procedure in 4-month-old children: although they may not show visible behavioral changes, both neurophysiology—vagal regulation—and the neuroendocrine system—the hypothalamic-pituitary-adrenal (HPA) axis—exhibit an initial adaptation to repeated exposure, 15 days later, to social stress in the laboratory (Montirosso et al., 2013; Montirosso et al., 2014).

Intentional

Between 2007 and 2014, over 140,000 elephants were killed by poaching. In Mozambique, the new generations of elephants are being born with a significant reduction or complete absence of tusks. This extreme adaptive process involves epigenetic mechanisms informed by the living environment of the previous generation and aims to ensure the survival of individuals at risk due to the behavior of other animals, humans being the adverse environment for these elephants. Preterm-born infants can gradually switch off genes involved in stress reactivity in response to a living environment rich in painful stimuli that could result in neurotoxic effects on the central nervous system due to complex hormonal cascades (Montirosso & Provenzi, 2015).

These examples illustrate how our biology makes us intentional beings, capable of making decisions within our environment and directing behavior in the physical and relational context in which we are immersed. However, it is crucial not to confuse intentionality with deliberation or finalism. These responses are often of a biological or physiological nature, largely influenced by ancestral response systems, such as the parasympathetic system or the HPA axis, that we share with many other animals and are the result of centuries of species evolution. Nevertheless, they are intentional movements that contribute silently but decisively to our life solutions. The extent to which these intentional biological learnings contribute to the developmental psychopathology is a question that is difficult to answer unequivocally. However, we can certainly imagine that these intentional biological learnings may exert long-term effects, influencing how we will engage with the physical and human environment in the future.

At the same time, the significant limitation of these intentional learnings is that our biology knows nothing of the future. Epigenetic or neuroendocrine intentional movements are responses based on an individual's history, the possibilities enabled by evolution; however, they cannot predict exactly how the emerging phenotype will be more or less coherent with the environmental conditions that the organism will encounter throughout its life. Biological learnings, therefore, involve compromises and bets on the relative stability of the living environment. The ability for a living system to maintain a reasonable degree of plasticity in the face of increasing stabilization of its organization is the real balance point on which the goodness of fit of its evolutionary trajectory depends (Chess & Thomas, 1999).

What's wrong?

The dominant model underlying much of the psychobiological research—even in the context of clinical and developmental psychology—views historical adaptations and responses of different neurobiological systems as acquired deficits in an individual's functioning based on encounters with adverse environments or particularly intense or prolonged stress conditions. Cumulative stress, toxic stress, the diathesis-stress model, and allostatic load model all these theoretical proposals are based on the concept of deficits, understood as the loss of certain functions or the efficiency of certain functions following early exposure to adverse contexts. In a sense, this approach leads to the question "What is wrong with you?" Consequently, interventions are configured as attempts to fix or recover what has been lost.

Of course, it is not a wrong story, some adverse environmental conditions can result in more or less marked deficits in the short or long term. And interventions developed based on a deficit theory have often proved effective. However, it is possible that the focus on deficits allows us to see only part of the story. When considering resilience, it is possible to consider the presence of individual differences such that some individuals even when exposed to particularly stressful contexts manage to cope and perhaps gain evolutionary benefits superior to individuals who have lived lives with fewer upheavals. The question posed by the resilience model is, therefore, "What made you successful?" In fact, resilience research has extensively focused on identifying protective elements or resilience factors—in biology, temperament traits, or certain environmental conditions.

Could we aspire to a better model? If we consider modifications of our biology following environmental exposures as life solutions—intentional positions or directions—we can better understand these same biological alterations as learnings rather than deficits. For example, Carolina de Weerth (Frankenhuis & de Weerth, 2013) has proposed to consider these adaptations as specializations: individuals exposed to a certain type of environment, which contributes to specific neurophysiological, psychobiological, and neuroendocrine adaptations, may become specialized individuals in providing a certain type of response or obtaining a certain type of reward in a similar context in the future.

In birds, individuals exposed to early stressful experiences show reduced growth, lower neuroendocrine regulation capabilities, reduced immune functionality, and suppression of some traits related to reproductive success, such as the ability to produce certain sounds (Crino & Breuner, 2015). However, these same birds may become particularly adept at coping with stress: they tend to be smaller, faster, and ultimately more capable of escaping predators. In the case of the zebra finch, exposure to stress may facilitate more rapid learning of foraging tasks (Crino et al., 2014). In the Japanese quail and domestic chicken, stress is associated with improved associative learning and spatial memory (Calandreau et al., 2011; Go-erlich et al., 2012). Rats exposed to early separation from their caregiver show increased hippocampal plasticity and learning abilities in future high-stress con-texts, such as contexts that elicit fear (Oomen et al., 2011). Furthermore, rodents raised by mothers that provide little licking-grooming are more capable of activat-ing defensive responses to a stranger (Menard & Hakvoort, 2007) and may be more sexually attractive to females (Sakhai et al., 2011). The potential advantages of learnings acquired in adverse contexts may be more evident when the animal's behavior is observed in the ecological environment, through a real-world approach. Using an operant learning task, Dunn et al. (2018) showed that individuals exposed to more stressful early life conditions displayed less weight but a greater ability to identify sources of food in contexts where food was harder to obtain. In the case of the zebra finch, chicks exposed to high levels of glucocorticoids learn how to pro-cure food by observing unfamiliar conspecifics' behavior instead of following their parents' behavior, suggesting that they are capable of deviating from evolution-imposed strategies to avoid reproducing potential errors from the previous genera-tion (Farine et al., 2015).

Pollak (2008) showed that children exposed to early abuse or maltreatment develop greater efficiency in identifying fearful faces, although in other studies, their ability to recognize faces displaying positive emotions is limited (Eisen et al., 2007). The behavioral response of preterm-born infants to socio-emotional stress shows higher displays of negative emotionality at 3 months compared to full-term infants of the same age (Montirosso et al., 2016b). This finding might reflect the preterm infants' need to "increase the volume" of their behavioral stress responses in the absence of adequate regulation support due to the partial presence of par-ents in the NICU; it is worth noting that the dissociation between behavioral and neurobiological stress responses might result in behavioral hyper-responsiveness, even in the presence of HPA axis hypo-reactivity (Montirosso et al., 2014; Provenzi et al., 2016d). Similarly, Mittal et al. (2015) suggested that individuals raised in unpredictable environments may develop the ability to quickly disengage attention to respond to sudden events, which facilitates the detection of threats and ena-bles the activation of faster and more functional response strategies. However, in a stable and predictable context, this function is less adaptive and can become an attentional problem. Thus, the researchers tested students with varying levels of exposure to stress and early adverse events on a control inhibitory attention task. The subjects were randomly assigned to a context in which the economic future was portrayed as unpredictable, characterized by limited resources and job loss, or to a control condition in which they were exposed to a series of relaxing images. In the control condition, no differences in inhibitory control emerged, while in the experimental condition, individuals who reported more frequent stressful events in their lives showed greater attentional shifting abilities compared to those who grew up in a more predictable and less problematic environment.

How do we proceed from here on out?

If we read the modifications and psychobiological deficits resulting from early experiences of stress not abstractly but in relation to the context, we can begin to understand these modifications as intentional biological learnings or attempts by the system to maximize complexity while maintaining adequate levels of coherence. Surviving and at the same time learning from our experiences. It is a gamble in which our biology seeks to benefit from our history and the present moment to program evolutionary trajectories that could be relatively adaptive in the near future. However, their adaptiveness largely depends on the context in which that developmental trajectory will continue to proceed.

In terms of scientific research, Ellis has recently initiated a study program called hidden talents aimed at understanding how early exposure to specific risk contexts can induce the emergence of hidden talents (Ellis et al., 2020). According to Ellis, the behavioral and neurobiological modifications that follow early adverse events can be considered as attempts at adaptation in that specific context, however unpredictable or adverse it may be. Specific adverse conditions can also shape developmental trajectories in very different ways, and the study of cumulative stress may not be able to clearly show the adaptive value of these modifications. Moreover, it is possible that adaptations occurring under risk conditions may—unexpectedly and surprisingly within the deficit model—lead to advantages in future contexts characterized by high competitiveness or limited resources (Frankenhuis et al., 2020).

I think of these hidden talents as a consequence of intentional biological learnings. It may be easier to observe these learnings in action by measuring any unexpected advantages in the adaptation of individuals exposed to early adverse conditions, by measuring the effect of specific environmental exposures, rather than vaguely defined conditions. In other words, studies investigating the effect of risk conditions, such as reduced socioeconomic status or poverty, may only partially capture the adaptive nature of biological learnings, as they would not allow observing the specific mechanisms stimulated by equally specific risk conditions. Similarly, it is likely easier to observe the functional consequences of intentional biological learnings by recording individuals' performance in their context rather than measuring personality traits with self-report instruments. High ecological validity studies are needed in the coming years to understand how to integrate the deficit model with a more respectful and capable vision of capturing the adaptive value of intentional biological learnings in adverse life contexts and at an early age. Compared to peers raised by highly sensitive mothers, rats raised by mothers showing reduced sensitivity show greater learning and memory abilities when tested under stressful conditions but not under baseline conditions (Champagne, 2010). Brazilian children raised in poverty conditions can solve complex mathematical tasks when tested in the same environment where they live and have learned to deal with clients in the market, but they perform worse at school (Schliemann & Carraher, 2002).

Reaching individuals at the point of their developmental trajectory, rather than asking them to approach a hypothetical trajectory of normality, should be the guiding principle of clinical action. In my training in psychodynamic therapy, I was taught to consider a patient's current state as the best possible configuration, the best life solution they could reach based on their history and resources. However, in the deficit model, this adaptive value risks being sidelined, and patients may experience shame precisely when seeking help (Finn, 2009). I believe this is one of the bravest challenges for clinical and developmental psychology today. If everything we are learning from developmental psychobiology can promote early interventions centered on the family—and therefore on the individual's living environment—and capable of optimizing the adaptiveness of intentional biological learnings, then it will have been much more successful than we could have imagined. Bruce Ellis (2012) argued that our corrective interventions are often highly dysfunctional because they try to teach a cat how to sheath its claws instead of how to use them to the best advantage.

In the coming years, we should become capable of developing interventions that do not start from the final goal or theory but from the exact point where an individual's life solutions—their intentional biological learnings—have brought them. Interventions capable of capitalizing on and adaptively using learnings accumulated during early life stages. The sooner we can inform our interventions based on the lessons of developmental psychobiology, the sooner we will be able to promote truly smart interventions, services, and policies, i.e., interventions that start from each individual's unique evolutionary trajectory to offer the best possible development conditions. Can we think of developing interventions capable of capitalizing on intentional biological learnings and teaching people how to use their own history in the present moment? I like to think that advances in developmental psychobiology can help us answer such an important question for both the scientific context and working with patients: "How do we proceed from here?"

Notes on contemporary issues, on frailty and care

Psychobiological postcards from the pandemic

I penned this book primarily during the spring of 2021, revisiting it a couple of years later to create the initial English version. The years in which the COVID-19 pandemic unfolded have defined this period of our lives, leaving an indelible mark and profoundly altering our existence—a transformation that continues as I conclude these reflections, contemplating the uncertain post-pandemic world. This unparalleled crisis has upheaved human life globally, exceeding our wildest imagination. Every facet of economic, social, and cultural life has been inundated. Some among us have bid farewell abruptly, often without the chance for a final embrace, to cherished ones. Many have grappled with job losses or faced substantial economic setbacks. Educational institutions, from schools to universities, have grappled with the challenges of adapting to remote learning, encountering numerous difficulties and organizational hurdles. Weeks and months of isolation have left many feeling alone. Science itself has encountered setbacks but is actively adapting. Simultaneously, numerous laboratories have collaborated to deliver a reliable, meticulous, yet swift and effective response to the urgent need for vaccines against a previously unknown virus.

In this scenario, there arises a question about the impact on our psychobiology. Considering the profound effects of significant events like the World Trade Center attack or the Dutch famine during World War II on those directly exposed and even on the second generation, it prompts us to contemplate the impact of the COVID-19 pandemic on our health. What outcomes should we anticipate? This is a query that warrants collaborative efforts from the realms of science and socio-health organizations. The mental well-being of those at the forefront of the pandemic—doctors, nurses, and healthcare personnel enduring months in the trenches—is facing severe challenges (Barello et al., 2020). Simultaneously, potential effects on the general population should not be underestimated. We were not prepared to confront an emergency condition that encapsulates many of the stressful dimensions studied in laboratory settings with animal models: social separations, resource deprivation, and unpredictability. The impact on the most vulnerable, those already grappling with life's challenges, could be immeasurable (Provenzi et al., 2020c; Shammi et al., 2020; van Dalen & Henkens, 2020).

The psychological aftermath of the pandemic is akin to a second tsunami, and the prolonged period of separation has left us yearning for physical touch (Banerjee et al., 2021). It is imperative that we commence addressing these consequences through targeted social, clinical, and policy interventions. While economic analyses of the damages will likely influence political decisions, scientific research has the potential and responsibility to contribute significantly. Developmental psychobiology can play a crucial role in delineating the parameters, mechanisms, outcomes, and potential areas for prevention related to various psychobiological processes implicated in stress during the pandemic. Examining the psychobiological implications of the ongoing socio-health emergency can aid in comprehending what could be characterized as a concealed pandemic, operating beneath the surface at the levels of our nervous, immune, and neuroendocrine systems. It represents a silent pandemic that may contribute to epigenetic regulations, forming biological memories of such a profound experience—a biological memory etched into the fabric of humanity.

Building on these reflections, in April 2019, I initiated a longitudinal and multicenter project titled "Measuring the Outcomes of Maternal COVID-19-related Prenatal Exposure," abbreviated as MOM-COPE (Provenzi et al., 2020d), at the Fondazione Mondino in Pavia. The primary objective was to comprehend how the stress endured by pregnant women during the pandemic months influences not only their psychological well-being but also the health of future children. In this project, behavioral measures are supplemented with other biological assessments, including the analysis of epigenetic changes in both the mother and the newborn.

Through this endeavor, we observed that elevated levels of prenatal stress correlated with heightened methylation levels of the SLC6A4 gene in the neonate, suggesting a biological transmission of stress from mother to fetus during pregnancy (Provenzi et al., 2021). These epigenetic modifications were subsequently associated with the child's temperament profile at 3 months of age (Provenzi et al., 2023) and socio-cognitive development at 12 months of age (Nazzari et al., 2023). It is crucial to highlight the temporal aspect of stress exposure and the potential significance of early intervention. Women exposed to lockdowns during the second and third trimesters of pregnancy had infants with more pronounced epigenetic alterations, indicating a potentially heightened sensitivity to this type of stress in the later stages of pregnancy (Nazzari et al., 2022). Conversely, families that received home visits during the postpartum period, even during lockdown phases, exhibited fewer anxious-depressive issues in women (Roberti et al., 2022). These findings underscore the importance of early support and intervention, even in normal circumstances, to facilitate the transition to parenthood and enhance the well-being of both parents and infants.

Thus, even amid the current emergency situation, a cohesive narrative emerges, enabling us to trace the imprints of early adverse experiences and comprehend their evolutionary consequences. In essence, these studies underscore our inherent fragility and vulnerability, emphasizing how our bodies consistently register our experiences, etching memories onto the pages of our DNA. These are inherent biological lessons that contribute to shaping our identity. Consequently, I believe we stand at a unique historical juncture to reassess our approach to care, shifting the focus away from the metaphor of battling pathology to acknowledging the imperative of human connections in our fragile state as human beings. By resolutely investing in human connections, we can foster greater well-being and development in preterm infants. Similarly, by addressing and mending the separations and disconnections experienced during the pandemic, we can glean invaluable biological insights silently unfolding within our bodies (Provenzi & Tronick, 2020).

In the current context, Europe is grappling with a resurgence of conflict in Ukraine. As a European citizen, the physical proximity magnifies the horrors of similar traumatic conditions in numerous other countries—ranging from Iran to Israel and Gaza, from Turkey to Sudan, to name a few—around the world. Looking forward, a significant challenge for developmental psychobiology is to establish translational partnerships with social, political, and health organizations, aiming to facilitate global sustainable development through informed policies and care interventions (see Sustainable Developmental Goals from United Nations). Key objectives involve promoting resilience, implementing trauma-informed care, encouraging early and effective interventions for children and their parents, preventing the intergenerational transmission of trauma, fostering healing and reconciliation, identifying and safeguarding vulnerable populations, reducing stigma, and advocating for social justice. These ethical avenues represent ways in which the developmental psychobiology of the near future can positively impact society on a global scale.

I strongly believe that in today's world, science has an obligation to effectively communicate not only its results but, crucially, its methodologies to civil society (Provenzi & Barello, 2020). The partial shortcomings of behavioral-based strategies for pandemic containment, exemplified by the limited success of tracking systems with applications, should not be solely attributed to citizens. Rather, there has been a deficiency in investing in a shared language capable of explaining, accompanying, and supporting individuals in their daily decision-making. Scientists must actively contribute to fostering a culture of scientific knowledge and promoting the virtuous use of communication media (Cox, 2020). Psychology, as the science of human behavior, should not shy away from a moral duty and has the opportunity to transcend the popular stereotype of the analyst and the chaise longue. Instead, it can genuinely evolve into a profession capable of initiating a revolution (Hillman and Ventura, 1998). By involving citizens in scientific endeavors, we can move beyond simplistic hero metaphors, which expose us to the extremes of a schizoid-paranoid interpretation of reality, and cultivate a culture of vulnerability and participatory civil responsibility. In the face of potential similar situations in the near or distant future, it would be gravely irresponsible to find ourselves unprepared.

Acknowledgments

This book represents the output of dynamic and collaborative efforts through space and time, involving numerous colleagues, friends, collaborators, and students.

I would like to extend special gratitude to Dr. Rosario Montirosso and the colleagues from the 0-3 Center at IRCCS E. Medea, where I had the privilege to conduct my research, to grow and develop as a human and as a researcher after completing my Master's in Psychology in 2008.

My sincere appreciation also goes to Renato Borgatti, who initially introduced me to the field of Infant Research and has been since the first time an exemplary figure in both professional and personal aspects throughout my career.

I am deeply thankful for the support and enthusiasm provided by my colleagues at the Developmental Psychobiology Lab, affiliated with the University of Pavia and IRCCS Mondino Foundation. Specifically, I am grateful to Elena Capelli, Serena Grumi, Sarah Nazzari, Miriam Pili, and Alessandra Raspanti for their valuable contributions.

Moreover, I am indebted to the students of the MSc program in Psychology, Neuroscience, and Human Sciences of the Department of Brain and Behavioral Science (University of Pavia), where I have the privilege of teaching Developmental Psychobiology and Developmental Psychopathology. Their engagement and interaction contributes to keep me curious, passionate, open to diversity and to continuous learning.

Special thanks to Elgin Bilge Tonga, who helped me in translating the book from its original Italian version. She was a Master student at the time of working on this volume.

This book, now in the hands of the readers, represents the culmination of intricate and profound collaborations with the individuals mentioned earlier and many others. These interwoven relationships have played a pivotal role in shaping the content and direction of this work. As readers journey through these pages, I am confident that they will share in my appreciation for the collective effort and contributions that have made this endeavor possible.

References

Aatsinki, A. K., Lahti, L., Uusitupa, H. M., Munukka, E., Keskitalo, A., Nolvi, S., & Karlsson, L. (2019). Gut microbiota composition is associated with temperament traits in infants. *Brain, Behavior, and Immunity, 80*, 849–858.

Åberg, E., Fandiño-Losada, A., Sjöholm, L. K., Forsell, Y., & Lavebratt, C. (2011). The functional Val158Met polymorphism in catechol-O-methyltransferase (COMT) is associated with depression and motivation in men from a Swedish population-based study. *Journal of Affective Disorders, 129*(1–3), 158–166.

Abraham, E., Hendler, T., Shapira-Lichter, I., Kanat-Maymon, Y., Zagoory-Sharon, O., & Feldman, R. (2014). Father's brain is sensitive to childcare experiences. *Proceedings of the National Academy of Sciences, 111*(27), 9792–9797.

Abrams, D. A., Chen, T., Odriozola, P., Cheng, K. M., Baker, A. E., Padmanabhan, A., Ryali, S., Kochalka, J., Feinstein, C., & Menon, V. (2016). Neural circuits underlying mother's voice perception predict social communication abilities in children. *Proceedings of the National Academy of Sciences of the United States of America, 113*(22), 6295–6300. https://doi.org/10.1073/pnas.1602948113

Acosta, J. K., & Levenson, R. L. Jr (2002). Observations from ground zero at the World Trade Center in New York City, part II: Theoretical and clinical considerations. *International Journal of Emergency Mental Health, 4*(2), 119–126.

Addabbo, M., Quadrelli, E., Bolognini, N., Nava, E., & Turati, C. (2020). Mirror-touch experiences in the infant brain. *Social Neuroscience, 15*(6), 641–649.

Afif, A., Bouvier, R., Buenerd, A., Trouillas, J., & Mertens, P. (2007). Development of the human fetal insular cortex: Study of the gyration from 13 to 28 gestational weeks. *Brain Structure and Function, 212*(3–4), 335–346.

Agostoni, E., Chinnock, J. E., Daly, M. D. B., & Murray, J. G. (1957). Functional and histological studies of the vagus nerve and its branches to the heart, lungs and abdominal viscera in the cat. *The Journal of Physiology, 135*(1), 182–205.

Ajilian Abbasi, M., Saeidi, M., Khademi, G., Hoseini, B. L., & Emami Moghadam, Z. (2015). Child maltreatment in the world: A review article. *International Journal of Pediatrics, 3*(1.1), 353–365.

Akther, S., Huang, Z., Liang, M., Zhong, J., Fakhrul, A. A., Yuhi, T., & Higashida, H. (2015). Paternal retrieval behavior regulated by brain estrogen synthetase (aromatase) in mouse sires that engage in communicative interactions with pairmates. *Frontiers in Neuroscience, 9*, 450.

Albers, E. M., Marianne Riksen-Walraven, J., Sweep, F. C., & Weerth, C. D. (2008). Maternal behavior predicts infant cortisol recovery from a mild everyday stressor. *Journal of Child Psychology and Psychiatry, 49*(1), 97–103.

Alexander, N., Kirschbaum, C., Wankerl, M., Stauch, B. J., Stalder, T., Steudte-Schmiedgen, S., & Miller, R. (2018). Glucocorticoid receptor gene methylation moderates the association of childhood trauma and cortisol stress reactivity. *Psychoneuroendocrinology, 90*, 68–75.

Alink, L. R., Cicchetti, D., Kim, J., & Rogosch, F. A. (2012). Longitudinal associations among child maltreatment, social functioning, and cortisol regulation. *Developmental Psychology*, *48*(1), 224–236.

Alkadhi, K. A. (2021). NMDA receptor-independent LTP in mammalian nervous system. *Progress in Neurobiology*, *200*, 101986.

Allman, J. M., Tetreault, N. A., Hakeem, A. Y., Manaye, K. F., Semendeferi, K., Erwin, J. M., Park, S., Goubert, V., & Hof, P. R. (2010). The von Economo neurons in frontoinsular and anterior cingulate cortex in great apes and humans. *Brain Structure and Function*, *214*(5–6), 495–517.

Allman, J. M., Watson, K. K., Tetreault, N. A., & Hakeem, A. Y. (2005). Intuition and autism: A possible role for von Economo neurons. *Trends in Cognitive Sciences*, *9*(8), 367–373.

Als, H., Duffy, F. H., McAnulty, G. B., Rivkin, M. J., Vajapeyam, S., Mulkern, R. V., & Eichenwald, E. C. (2004). Early experience alters brain function and structure. *Pediatrics*, *113*(4), 846–857.

Altuncu, E., Akman, I., Kulekci, S., Akdas, F., Bilgen, H., & Ozek, E. (2009). Noise levels in neonatal intensive care unit and use of sound absorbing panel in the isolette. *International Journal of Pediatric Otorhinolaryngology*, *73*(7), 951–953.

Ancharoff, M. R., Munroe, J. F., & Fisher, L. M. (1998). The legacy of combat trauma. In Y. Danieli (Ed.), (a cura di), *International handbook of multigenerational legacies of trauma* (pp. 257–276). Boston: Springer.

Andrews, M. H., & Matthews, S. G. (2004). Programming of the hypothalamo–pituitary–adrenal axis: Serotonergic involvement. *Stress*, *7*(1), 15–27.

Appleton, A. A., Lester, B. M., Armstrong, D. A., Lesseur, C., & Marsit, C. J. (2015). Examining the joint contribution of placental NR3C1 and HSD11B2 methylation for infant neurobehavior. *Psychoneuroendocrinology*, *52*, 32–42.

Arabadzisz, D., Diaz-Heijtz, R., Knuesel, I., Weber, E., Pilloud, S., Dettling, A. C., & Pryce, C. R. (2010). Primate early life stress leads to long-term mild hippocampal decreases in corticosteroid receptor expression. *Biological Psychiatry*, *67*(11), 1106–1109.

Armbruster, D., Mueller, A., Strobel, A., Lesch, K. P., Brocke, B., & Kirschbaum, C. (2012). Children under stress–COMT genotype and stressful life events predict cortisol increase in an acute social stress paradigm. *International Journal of Neuropsychopharmacology*, *15*(9), 1229–1239.

Arslan, A. A., Tuminello, S., Yang, L., Zhang, Y., Durmus, N., Snuderl, M., & Reibman, J. (2020). Genome-wide DNA methylation profiles in community members exposed to the World Trade Center disaster. *International Journal of Environmental Research and Public Health*, *17*(15), 5493.

Assoun, P. L. (1988). *Introduzione all'epistemologia freudiana*. Theoria.

Atzil, S., Hendler, T., & Feldman, R. (2011). Specifying the neurobiological basis of human attachment: Brain, hormones, and behavior in synchronous and intrusive mothers. *Neuropsychopharmacology*, *36*(13), 2603–2615.

Atzil, S., Hendler, T., Zagoory-Sharon, O., Winetraub, Y., & Feldman, R. (2012). Synchrony and specificity in the maternal and the paternal brain: Relations to oxytocin and vasopressin. *Journal of the American Academy of Child Adolescent Psychiatry*, *51*(8), 798–811.

Auerbach, J., Geller, V., Lezer, S., Shinwell, E., Belmaker, R. H., Levine, J., & Ebstein, R. P. (1999). Dopamine D4 receptor (D4DR) and serotonin transporter promoter (5-HTTLPR) polymorphisms in the determination of temperament in 2-month-old infants. *Molecular Psychiatry*, *4*(4), 369–373.

Austerberry, C., & Fearon, P. (2020). An overview of developmental behavioral genetics. In: Provenzi, L., Montirosso, R. (a cura di), *Developmental human behavioral epigenetics*, 59–80. Academic Press.

Aviv, A., Hunt, S. C., Lin, J., Cao, X., Kimura, M., & Blackburn, E. (2011). Impartial comparative analysis of measurement of leukocyte telomere length/DNA content by Southern blots and qPCR. *Nucleic Acids Research*, *39*(20), e134–e134.

Azarian-Ceccato, N. (2010). Reverberations of the Armenian genocide: Narrative's intergenerational transmission and the task of not forgetting. *Narrative Inquiry, 20*(1), 106–123.

Azhari, A., Bizzego, A., & Esposito, G. (2020). Father-child dyads exhibit unique intersubject synchronisation during coviewing of animation video stimuli. BioRxiv preprint, https://doi.org/10.1101/2020.10.30.361592;

Azhari, A., Leck, W. Q., Gabrieli, G., Bizzego, A., Rigo, P., Setoh, P., & Esposito, G. (2019). Parenting stress undermines mother-child brain-to-brain synchrony: A hyperscanning study. *Scientific Reports, 9*(1), 1–9.

Bader, L. R., Tan, L., Gonzalez, R., Saini, E. K., Bae, Y., Provenzi, L., & Volling, B. L. (2021). Adrenocortical interdependence in father-infant and mother-infant dyads: Attunement or something more? *Developmental Psychobiology, 63*(5), 1534–1548.

Bakermans-Kranenburg, M. J., & van Ijzendoorn, M. H. (2006). Gene-environment interaction of the dopamine D4 receptor (DRD4) and observed maternal insensitivity predicting externalizing behavior in preschoolers. *Developmental Psychobiology, 48*(5), 406–409.

Bakermans-Kranenburg, M. J., van Ijzendoorn, M. H., Pijlman, F. T., Mesman, J., & Juffer, F. (2008). Experimental evidence for differential susceptibility: Dopamine D4 receptor polymorphism (DRD4 VNTR) moderates intervention effects on toddlers' externalizing behavior in a randomized controlled trial. *Developmental Psychology, 44*(1), 293.

Baltimore, D. (2001). Our genome unveiled. *Nature, 409*(6822), 815–816.

Bambico, F. R., Lacoste, B., Hattan, P. R., & Gobbi, G. (2015). Father absence in the monogamous California mouse impairs social behavior and modifies dopamine and glutamate synapses in the medial prefrontal cortex. *Cerebral Cortex, 25*(5), 1163–1175.

Banerjee, D., Vasquez, V., Pecchio, M., Hegde, M. L., Ks Jagannatha, R., & Rao, T. S. (2021). Biopsychosocial intersections of social/affective touch and psychiatry: Implications of 'touch hunger' during COVID-19. *The International Journal of Social Psychiatry*, 20764021997485.

Banning, C. (1946). Food shortage and public health, first half of 1945. *The ANNALS of the American Academy of Political and Social Science, 245*(1), 93–110.

Barello, S., Palamenghi, L., & Graffigna, G. (2020). Burnout and somatic symptoms among frontline healthcare professionals at the peak of the Italian COVID-19 pandemic. *Psychiatry Research, 290*, 113129.

Baricco, A. (1991). *Castelli di rabbia*. Feltrinelli: Milano.

Baricco, A. (1994). *Novecento*. Feltrinelli: Milano.

Barker, D. J. P. (2004). The developmental origins of adult disease. *Journal of the American College of Nutrition, 23*(sup6), 588S–595S.

Barker, D. J., & Osmond, C. (1986). Diet and coronary heart disease in England and Wales during and after the second world war. *Journal of Epidemiology and Community Health, 40*(1), 37–44.

Barker, D. J., Osmond, C., Winter, P. D., Margetts, B., & Simmonds, S. J. (1989). Weight in infancy and death from ischaemic heart disease. *The Lancet, 334*(8663), 577–580.

Barriga-Vallejo, C., Hernandez-Gallegos, O., Von Herbing, I. H., López-Moreno, A. E., Ruiz-Gómez, M. D. L., Granados-Gonzalez, G., & Davis, A. K. (2015). Assessing population health of the Toluca axolotl *Ambystoma rivulare* (Taylor, 1940) from Mexico, using leukocyte profiles. *Herpetological Conservation and Biology, 10*(2), 592–601.

Barr, C. S., Newman, T. K., Shannon, C., Parker, C., Dvoskin, R. L., Becker, M. L., & Higley, J. D. (2004). Rearing condition and rh5-HTTLPR interact to influence limbic-hypothalamic-pituitary-adrenal axis response to stress in infant macaques. *Biological Psychiatry, 55*(7), 733–738.

Barron, I. G., & Abdallah, G. (2015). Intergenerational trauma in the occupied Palestinian territories: Effect on children and promotion of healing. *Journal of Child & Adolescent Trauma, 8*(2), 103–110.

Barry, R. A., Kochanska, G., & Philibert, R. A. (2008). G x E interaction in the organization of attachment: mothers' responsiveness as a moderator of children's genotypes. *Journal of Child Psychology and Psychiatry, and Allied Disciplines, 49*(12), 1313–1320.

Bartels, A., & Zeki, S. (2004). The neural correlates of maternal and romantic love. *Neuro-Image, 21*(3), 1155–1166.

Bates, M. G., Newman, J. H., Jakovljevic, D. G., Hollingsworth, K. G., Alston, C. L., Zalewski, P., & Gorman, G. S. (2013). Defining cardiac adaptations and safety of endurance training in patients with m. 3243A> G-related mitochondrial disease. *International Journal of Cardiology, 168*(4), 3599–3608.

Bayerl, D. S., Hönig, J. N., & Bosch, O. J. (2016). Vasopressin V1a, but not V1b, receptors within the PVN of lactating rats mediate maternal care and anxiety-related behaviour. *Behavioural Brain Research, 305*, 18–22.

Bazhenova, O. V., Plonskaia, O., & Porges, S. W. (2001). Vagal reactivity and affective adjustment in infants during interaction challenges. *Child Development, 72*(5), 1314–1326.

Beach, S. R., Dogan, M. V., Brody, G. H., & Philibert, R. A. (2014). Differential impact of cumulative SES risk on methylation of protein–protein interaction pathways as a function of SLC6A4 genetic variation in African American young adults. *Biological Psychology, 96*, 28–34.

Beauchaine, T. P., Bell, Z., Knapton, E., McDonough-Caplan, H., Shader, T., & Zisner, A. (2019). Respiratory sinus arrhythmia reactivity across empirically based structural dimensions of psychopathology: A meta-analysis. *Psychophysiology, 56*(5), e13329.

Beauchaine, T. P., Gatzke-Kopp, L., Neuhaus, E., Chipman, J., Reid, M. J., & Webster-Stratton, C. (2013). Sympathetic-and parasympathetic-linked cardiac function and prediction of externalizing behavior, emotion regulation, and prosocial behavior among preschoolers treated for ADHD. *Journal of Consulting and Clinical Psychology, 81*(3), 481.

Beebe, B. E., & Lachmann, F. M. (2005). *Infant research and Adult Treatment.* New York, NY: Routledge.

Beers, S. R., & De Bellis, M. D. (2002). Neuropsychological function in children with maltreatment-related posttraumatic stress disorder. *American Journal of Psychiatry, 159*(3), 483–486.

Bellugi, U., Wang, P. P., & Jernigan, T. L. (1994). Williams syndrome: An unusual neuropsychological profile. *Atypical Cognitive Deficits in Developmental Disorders: Implications for Brain Function, 23*, 23–56.

Belsky, J. (1997). Theory testing, effect-size evaluation, and differential susceptibility to rearing influence: The case of mothering and attachment. *Child Development, 68*(4), 598–600.

Belsky, J. (2002). Developmental origins of attachment styles. *Attachment Human Development, 4*(2), 166–170.

Belsky, J., Bakermans-Kranenburg, M. J., & van Ijzendoorn, M. H. (2007). For better and for worse: Differential susceptibility to environmental influences. *Current Directions in Psychological Science, 16*(6), 300–304.

Belsky, J. A. Y., Hsieh, K. H., & Crnic, K. (1998). Mothering, fathering, and infant negativity as antecedents of boys' externalizing problems and inhibition at age 3 years: Differential susceptibility to rearing experience? *Development and Psychopathology, 10*(2), 301–319.

Belsky, J., & Pluess, M. (2009). The nature (and nurture?) Of plasticity in early human development. *Perspectives on Psychological Science, 4*(4), 345–351.

Belsky, J., Steinberg, L., & Draper, P. (1991). Childhood experience, interpersonal development, and reproductive strategy: An evolutionary theory of socialization. *Child Development, 62*(4), 647–670.

Ben Shalom, D., Mostofsky, S. H., Hazlett, R. L., Goldberg, M. C., Landa, R. J., Faran, Y., McLeod, D. R., & Hoehn-Saric, R. (2006). Normal physiological emotions but differences in expression of conscious feelings in children with high-functioning autism. *Journal of Autism and Developmental Disorders, 36*(3), 395–400.

Bendesky, A., Kwon, Y. M., Lassance, J. M., Lewarch, C. L., Yao, S., Peterson, B. K., & Hoekstra, H. E. (2017). The genetic basis of parental care evolution in monogamous mice. *Nature, 544*(7651), 434–439.

Bernard, C. (1865). Etude sur la physiologie du coeur. *Revue Des Deux Mondes (1829-1971), 56*(1), 236–252.

Bernard, C. (1879). Leçons sur les phénomènes de la vie commune aux animaux et aux végétaux (Vol. 2). Baillière.

Bernard, N. K., Kashy, D. A., Levendosky, A. A., Bogat, G. A., & Lonstein, J. S. (2017). Do different data analytic approaches generate discrepant findings when measuring mother–infant HPA axis attunement? *Developmental Psychobiology, 59*(2), 174–184.

Berretta, E., Cutuli, D., Laricchiuta, D., Petrosini, L. (2020). From animal to human epigenetics. In: Provenzi, L., Montirosso, R. (a cura di), Developmental human behavioral epigenetics (pp. 27–58). Academic Press.

Berretta, E., Guida, E., Forni, D., & Provenzi, L. (2021). Glucocorticoid receptor gene (NR3C1) methylation during the first 1000 days: Environmental exposures and developmental outcomes. *Neuroscience Biobehavioral Reviews, 125*, 493–502.

Berthoz, S., Artiges, E., Van De Moortele, P. F., Poline, J. B., Rouquette, S., Consoli, S. M., & Martinot, J. L. (2002). Effect of impaired recognition and expression of emotions on frontocingulate cortices: an fMRI study of men with alexithymia. *The American journal of psychiatry, 159*(6), 961–967. https://doi.org/10.1176/appi.ajp.159.6.961

Bertone-Johnson, E. R., Whitcomb, B. W., Missmer, S. A., Karlson, E. W., & Rich-Edwards, J. W. (2012). Inflammation and early-life abuse in women. *American Journal of Preventive Medicine, 43*(6), 611–620.

Bestor, T. H., & Ingram, V. M. (1983). Two DNA methyltransferases from murine erythroleukemia cells: Purification, sequence specificity, and mode of interaction with DNA. *Proceedings of the National Academy of Sciences, 80*(18), 5559–5563.

Betz, V. (1881). Sur la structure de l'écorce cérébrale. *Revue d'Anthropologie, 3*, 426–428.

Bevans, K., Cerbone, A., & Overstreet, S. (2008). Relations between recurrent trauma exposure and recent life stress and salivary cortisol among children. *Development and Psychopathology, 20*(1), 257–272.

Bharwani, A., Mian, M. F., Foster, J. A., Surette, M. G., Bienenstock, J., & Forsythe, P. (2016). Structural functional consequences of chronic psychosocial stress on the microbiome host. *Psychoneuroendocrinology, 63*, 217–227.

Biaggi, A., Conroy, S., Pawlby, S., & Pariante, C. M. (2016). Identifying the women at risk of antenatal anxiety and depression: A systematic review. *Journal of Affective Disorders, 191*, 62–77.

Bierhaus, A., Wolf, J., Andrassy, M., Rohleder, N., Humpert, P. M., Petrov, D., & Nawroth, P. P. (2003). A mechanism converting psychosocial stress into mononuclear cell activation. *Proceedings of the National Academy of Sciences, 100*(4), 1920–1925.

Bimbi, M., Festante, F., Coudé, G., Vanderwert, R. E., Fox, N. A., & Ferrari, P. F. (2018). Simultaneous scalp recorded EEG and local field potentials from monkey ventral premotor cortex during action observation and execution reveals the contribution of mirror and motor neurons to the mu-rhythm. *NeuroImage, 175*, 22–31.

Birnie, A. K., Taylor, J. H., Cavanaugh, J., & French, J. A. (2013). Quality of maternal and paternal care predicts later stress reactivity in the cooperatively-breeding marmoset (*Callithrix geoffroyi*). *Psychoneuroendocrinology, 38*(12), 3003–3014.

Blackburn, E. H. (1990). Telomeres and their synthesis. *Science, 249*(4968), 489–491.

Bliss, T. V., & Lømo, T. (1973). Long-lasting potentiation of synaptic transmission in the dentate area of the anaesthetized rabbit following stimulation of the perforant path. *The Journal of Physiology, 232*(2), 331–356.

Blum, K., Noble, E. P., Sheridan, P. J., Montgomery, A., Ritchie, T., Jagadeeswaran, P., & Cohn, J. B. (1990). Allelic association of human dopamine D2 receptor gene in alcoholism. *JAMA, 263*(15), 2055–2060.

Boeck, C., Koenig, A. M., Schury, K., Geiger, M. L., Karabatsiakis, A., Wilker, S., & Kolassa, I. T. (2016). Inflammation in adult women with a history of child maltreatment: The involvement of mitochondrial alterations and oxidative stress. *Mitochondrion, 30*, 197–207.

Booij, L., Wang, D., Lévesque, M. L., Tremblay, R. E., & Szyf, M. (2013). Looking beyond The DNA sequence: The relevance of DNA methylation processes for The stress–diathesis model of depression. *Philosophical Transactions of the Royal Society B: Biological Sciences*, 368(1615), 20120251.

Bosch, O. J., & Neumann, I. D. (2012). Both oxytocin and vasopressin are mediators of maternal care and aggression in rodents: From Central release to sites of action. *Hormones and Behavior, 61*(3), 293–303.

Boston Change Process Study Group (2012). *Il cambiamento in psicoterapia*. Milano: Raffaello Cortina Editore.

Boyapati, R. K., Tamborska, A., Dorward, D. A., Ho, G. T. (2017). Advances in the understanding of mitochondrial DNA as a pathogenic factor in inflammatory diseases. F1000Research, 6.

Boyce, W. T., Chesney, M., Alkon, A., Tschann, J. M., Adams, S., Chesterman, B., & Wara, D. (1995). Psychobiologic reactivity to stress and childhood respiratory illnesses: Results of two prospective studies. *Psychosomatic Medicine, 57*(5), 411–422.

Boyce, W. T., & Ellis, B. J. (2005). Biological sensitivity to context: I. An evolutionary–developmental theory of the origins and functions of stress reactivity. *Development and Psychopathology, 17*(2), 271–301.

Bradley, R. H., & Corwyn, R. F. (2008). Infant temperament, parenting, and externalizing behavior in first grade: A test of the differential susceptibility hypothesis. *Journal of Child Psychology and Psychiatry, 49*(2), 124–131.

Bradley, K., & Eccles, J. C. (1953). Analysis of the fast afferent impulses from thigh muscles. *The Journal of physiology, 122*(3), 462.

Braithwaite, E. C., Kundakovic, M., Ramchandani, P. G., Murphy, S. E., & Champagne, F. A. (2015). Maternal prenatal depressive symptoms predict infant NR3C1 1F and BDNF IV DNA methylation. *Epigenetics, 10*(5), 408–417.

Bredy, T. W., Brown, R. E., & Meaney, M. J. (2007). Effect of resource availability on biparental care, and offspring neural and behavioral development in the California mouse (peromyscus californicus). *European Journal of Neuroscience, 25*(2), 567–575.

Brenna, V., Proietti, V., Montirosso, R., & Turati, C. (2013). Positive, but not negative, facial expressions facilitate 3-month-olds' recognition of an individual face. International Journal of Behavioral Development, 37(2), 137–142. https://doi.org/10.1177/0165025412465363

Bridges, R. S. (2015). Neuroendocrine regulation of maternal behavior. *Frontiers in Neuroendocrinology, 36*, 178–196.

Bromer, C., Marsit, C. J., Armstrong, D. A., Padbury, J. F., & Lester, B. (2013). Genetic and epigenetic variation of the glucocorticoid receptor (NR3C1) in placenta and infant neurobehavior. *Developmental Psychobiology, 55*(7), 673–683.

Brown, G. (2009). NICU noise and the preterm infant. *Neonatal Network, 28*(3), 165–173.

Brummelte, S., Grunau, R. E., Synnes, A. R., Whitfield, M. F., & Petrie-Thomas, J. (2011). Declining cognitive development from 8 to 18 months in preterm children predicts persisting higher parenting stress. *Early Human Development, 87*(4), 273–280.

Brummelte, S., Pawluski, J. L., & Galea, L. A. (2006). High post-partum levels of corticosterone given to dams influence postnatal hippocampal cell proliferation and behavior of offspring: A model of post-partum stress and possible depression. *Hormones and Behavior, 50*(3), 370–382.

Brydon, L., Edwards, S., Jia, H., Mohamed-Ali, V., Zachary, I., Martin, J. F., & Steptoe, A. (2005). Psychological stress activates interleukin-1β gene expression in human mononuclear cells. *Brain, Behavior, and Immunity, 19*(6), 540–546.

Buchanan, T. W., Tranel, D., & Adolphs, R. (2006). Impaired memory retrieval correlates with individual differences in cortisol response but not autonomic response. *Learning Memory, 13*(3), 382–387.

Burke, S. (2018). Systematic review of developmental care interventions in the neonatal intensive care unit since 2006. *Journal of Child Health Care, 22*(2), 269–286.

Buss, C., Entringer, S., Reyes, J. F., Chicz-Demet, A., Sandman, C. A., Waffarn, F., & Wadhwa, P. D. (2009). The maternal cortisol awakening response in human pregnancy is associated with the length of gestation. *AJOG, 201*(4), 398–e1.

Bustamante, A. C., Aiello, A. E., Galea, S., Ratanatharathorn, A., Noronha, C., Wildman, D. E., & Uddin, M. (2016). Glucocorticoid receptor DNA methylation, childhood maltreatment and major depression. *Journal of Affective Disorders, 206*, 181–188.

Butler, S. C., O'sullivan, L. P., Shah, B. L., & Berthier, N. E. (2014). Preference for infant-directed speech in preterm infants. *Infant Behavior and Development, 37*(4), 505–511.

Butti, C., Raghanti, M. A., Sherwood, C. C., & Hof, P. R. (2011). The neocortex of cetaceans: cytoarchitecture and comparison with other aquatic and terrestrial species. *Annals of the New York Academy of Sciences, 1225*, 47–58. https://doi.org/10.1111/j.1749-6632.2011.05980.x

Butti, C., Sherwood, C. C., Hakeem, A. Y., Allman, J. M., & Hof, P. R. (2009). Total number and volume of von economo neurons in the cerebral cortex of cetaceans. *The Journal of Comparative Neurology, 515*(2), 243–259.

Byrnes, E. M., Casey, K., Carini, L. M., & Bridges, R. S. (2013). Reproductive experience alters neural and behavioural responses to acute oestrogen receptor α activation. *Journal of Neuroendocrinology, 25*(12), 1280–1289.

Byrne, R. W., & Whiten, A. (1988). *Machiavellian intelligence: Social expertise and the evolution of intellect in monkeys, apes, and humans.* Clarendon Press/Oxford University Press.

Calandreau, L., Favreau-Peigné, A., Bertin, A., Constantin, P., Arnould, C., Laurence, A., & Leterrier, C. (2011). Higher inherent fearfulness potentiates the effects of chronic stress in the Japanese quail. *Behavioural Brain Research, 225*(2), 505–510.

Calciolari, G., & Montirosso, R. (2011). The sleep protection in the preterm infants. *The Journal of Maternal-Fetal Neonatal Medicine, 24*(sup1), 12–14.

Caldji, C., Tannenbaum, B., Sharma, S., Francis, D., Plotsky, P. M., & Meaney, M. J. (1998). Maternal care during infancy regulates the development of neural systems mediating the expression of fearfulness in the rat. *Proceedings of the National Academy of Sciences, 95*(9), 5335–5340.

Canetti, D., Russ, E., Luborsky, J., Gerhart, J. I., & Hobfoll, S. E. (2014). Inflamed by the flames? The impact of terrorism and war on immunity. *Journal of Traumatic Stress, 27*(3), 345–352.

Canli, T., & Lesch, K. P. (2007). Long story short: The serotonin transporter in emotion regulation and social cognition. *Nature Neuroscience, 10*(9), 1103–1109.

Cannon, W. B. (1929). Organization for physiological homeostasis. *Physiological Reviews, 9*(3), 399–431.

Cao-Lei, L., Massart, R., Suderman, M. J., Machnes, Z., Elgbeili, G., Laplante, D. P., & King, S. (2014). DNA methylation signatures triggered by prenatal maternal stress exposure to a natural disaster: Project Ice Storm. *PLoS One, 9*(9), e107653.

Cao-Lei, L., Veru, F., Elgbeili, G., Szyf, M., Laplante, D. P., & King, S. (2016). DNA methylation mediates the effect of exposure to prenatal maternal stress on cytokine production in children at age 13½ years: Project ice storm. *Clinical Epigenetics, 8*(1), 1–15.

Carlson, M., & Earls, F. (1997). Psychological and neuroendocrinological sequelae of early social deprivation in institutionalized children in Romania. *Annals of the New York Academy of Sciences, 807*(1), 419–428.

Carlson, A. L., Xia, K., Azcarate-Peril, M. A., Goldman, B. D., Ahn, M., Styner, M. A., & Knickmeyer, R. C. (2018). Infant gut microbiome associated with cognitive development. *Biological Psychiatry, 83*(2), 148–159.

Carpenter, L. L., Carvalho, J. P., Tyrka, A. R., Wier, L. M., Mello, A. F., Mello, M. F., & Price, L. H. (2007). Decreased adrenocorticotropic hormone and cortisol responses to stress in healthy adults reporting significant childhood maltreatment. *Biological Psychiatry, 62*(10), 1080–1087.

Carpenter, L. L., Gawuga, C. E., Tyrka, A. R., Lee, J. K., Anderson, G. M., & Price, L. H. (2010). Association between plasma IL-6 response to acute stress and early-life adversity in healthy adults. *Neuropsychopharmacology, 35*(13), 2617–2623.

Carpenter, L. L., Shattuck, T. T., Tyrka, A. R., Geracioti, T. D., & Price, L. H. (2011). Effect of childhood physical abuse on cortisol stress response. *Psychopharmacology, 214*(1), 367–375.

Carter, C. S. (2014). Oxytocin pathways and the evolution of human behavior. *Annual Review of Psychology, 65*, 17–39.

Casey, B. J., Galvan, A., & Hare, T. A. (2005). Changes in cerebral functional organization during cognitive development. *Current Opinion in Neurobiology, 15*(2), 239–244.

Caskey, M., Stephens, B., Tucker, R., & Vohr, B. (2014). Adult talk in the NICU with preterm infants and developmental outcomes. *Pediatrics, 133*(3), e578–e584.

Caso, J. R., Moro, M. A., Lorenzo, P., Lizasoain, I., & Leza, J. C. (2007). Involvement of IL-1β in acute stress-induced worsening of cerebral ischaemia in rats. *European Neuropsychopharmacology, 17*(9), 600–607.

Caspi, A., McClay, J., Moffitt, T. E., Mill, J., Martin, J., Craig, I. W., & Poulton, R. (2002). Role of genotype in the cycle of violence in maltreated children. *Science, 297*(5582), 851–854.

Caspi, A., Sugden, K., Moffitt, T. E., Taylor, A., Craig, I. W., Harrington, H., & Poulton, R. (2003). Influence of life stress on depression: Moderation by a polymorphism in the 5-HTT gene. *Science, 301*(5631), 386–389.

Cattaneo, A., Suderman, M., Cattane, N., Mazzelli, M., Begni, V., Maj, C., & Riva, M. A. (2020). Long-term effects of stress early in life on microRNA-30a and its network: Preventive effects of lurasidone and potential implications for depression vulnerability. *Neurobiology of Stress, 13*, 100271.

Cawthon, R. M. (2002). Telomere measurement by quantitative PCR. *Nucleic Acids Research, 30*(10), e47–e47.

Cevasco, A. M. (2008). The effects of mothers' singing on full-term and preterm infants and maternal emotional responses. *Journal of Music Therapy, 45*(3), 273–306.

Champagne, F. A. (2008). Epigenetic mechanisms and the transgenerational effects of maternal care. *Frontiers in Neuroendocrinology, 29*(3), 386–397.

Champagne, F. A. (2010). Epigenetic influence of social experiences across the lifespan. *Developmental psychobiology, 52*(4), 299–311.

Champagne, F., Diorio, J., Sharma, S., & Meaney, M. J. (2001). Naturally occurring variations in maternal behavior in the rat are associated with differences in estrogen-inducible central oxytocin receptors. *Proceedings of the National Academy of Sciences, 98*(22), 12736–12741.

Champagne, F. A., Francis, D. D., Mar, A., & Meaney, M. J. (2003). Variations in maternal care in the rat as a mediating influence for the effects of environment on development. *Physiology Behavior, 79*(3), 359–371.

Changeux, J. P., Galzi, J. L., Devillers-Thiéry, A., & Bertrand, D. (1992). The functional architecture of the acetylcholine nicotinic receptor explored by affinity labelling and site-directed mutagenesis. *Quarterly Reviews of Biophysics, 25*(04), 395.

Charbonneau, M. R., Blanton, L. V., Digiulio, D. B., Relman, D. A., Lebrilla, C. B., Mills, D. A., & Gordon, J. I. (2016). A microbial perspective of human developmental biology. *Nature, 535*(7610), 48–55.

Chargaff, E. (1950). Chemical specificity of nucleic acids and mechanism of their enzymatic degradation. *Experientia, 6*(6), 201–209.

Chau, C. M., Ranger, M., Sulistyoningrum, D., Devlin, A. M., Oberlander, T. F., & Grunau, R. E. (2014). Neonatal pain and COMT Val158Met genotype in relation to serotonin transporter (SLC6A4) promoter methylation in very preterm children at school age. *Frontiers in Behavioral Neuroscience, 8*, 409.

Chen, J., Lipska, B. K., Halim, N., Ma, Q. D., Matsumoto, M., Melhem, S., & Weinberger, D. R. (2004). Functional analysis of genetic variation in catechol-O-methyltransferase (COMT): Effects on mRNA, protein, and enzyme activity in postmortem human brain. *The American Journal of Human Genetics, 75*(5), 807–821.

Chess, S., & Thomas, A. (1999). *Goodness of fit: Clinical applications from infancy through adult life.* Psychology Press.

Chorna, O. D., Slaughter, J. C., Wang, L., Stark, A. R., & Maitre, N. L. (2014). A pacifier-activated music player with mother's voice improves oral feeding in preterm infants. *Pediatrics, 133*(3), 462–468.

Cho, I., Yamanishi, S., Cox, L., Methé, B. A., Zavadil, J., Li, K., & Blaser, M. J. (2012). Antibiotics in early life alter the murine colonic microbiome and adiposity. *Nature, 488*(7413), 621–626.

Chrousos, G. P., Loriaux, D. L., & Gold, P. W. (1988). Introduction: The concept of stress and its historical development. *Mechanisms of physical and emotional stress*, 3–7. Springer: Boston.

Chu, C., Murdock, M. H., Jing, D., Won, T. H., Chung, H., Kressel, A. M., & Artis, D. (2019). The microbiota regulate neuronal function and fear extinction learning. *Nature, 574*(7779), 543–548.

Cicchetti, D., & Handley, E. D. (2017). Methylation of the glucocorticoid receptor gene, nuclear receptor subfamily 3, group C, member 1 (NR3C1), in maltreated and nonmaltreated children: Associations with behavioral undercontrol, emotional lability/negativity, and externalizing and internalizing symptoms. *Development and Psychopathology, 29*(5), 1795–1806.

Cicchetti, D., Rogosch, F. A., Gunnar, M. R., & Toth, S. L. (2010). The differential impacts of early physical and sexual abuse and internalizing problems on daytime cortisol rhythm in school-aged children. *Child Development, 81*(1), 252–269.

Cicchetti, D., Rogosch, F. A., Lynch, M., & Holt, K. D. (1993). Resilience in maltreated children: Processes leading to adaptive outcome. *Development and psychopathology, 5*(4), 629–647.

Clarke, G., Grenham, S., Scully, P., Fitzgerald, P., Moloney, R. T., Shanahan, F., & Cryan, J. T. (2013). The microbiome-gut-brain axis during early life regulates the hippocampal serotonergic system in a sex-dependent manner. *Molecular Psychiatry, 18*(6), 666–673.

Coan, J. A., & Allen, J. J. (2004). Frontal EEG asymmetry as a moderator and mediator of emotion. *Biological Psychology, 67*(1–2), 7–50.

Cohen-Bendahan, C. C., Beijers, R., van Doornen, L. J., & de Weerth, C. (2015). Explicit and implicit caregiving interests in expectant fathers: Do endogenous and exogenous oxytocin and vasopressin matter? *Infant Behavior and Development, 41*, 26–37.

Cong, X., Ludington-Hoe, S. M., Hussain, N., Cusson, R. M., Walsh, S., Vazquez, V., & Vittner, D. (2015). Parental oxytocin responses during skin-to-skin contact in pre-term infants. *Early Human Development, 91*(7), 401–406.

Conradt, E., Adkins, D. E., Crowell, S. E., Raby, K. L., Diamond, L. M., & Ellis, B. (2018). Incorporating epigenetic mechanisms to advance fetal programming theories. *Development and Psychopathology, 30*(3), 807–824.

Conradt, E., Hawes, K., Guerin, D., Armstrong, D. A., Marsit, C. J., Tronick, E., & Lester, B. M. (2016). The contributions of maternal sensitivity and maternal depressive symptoms to epigenetic processes and neuroendocrine functioning. *Child Development, 87*(1), 73–85.

Conradt, E., Lester, B. M., Appleton, A. A., Armstrong, D. A., & Marsit, C. J. (2013). The roles of DNA methylation of NR3C1 and 11β-HSD2 and exposure to maternal mood disorder in utero on newborn neurobehavior. *Epigenetics, 8*(12), 1321–1329.

Conradt, E., Ostlund, B., Guerin, D., Armstrong, D. A., Marsit, C. J., Tronick, E., & Lester, B. M. (2019). DNA methylation of NR3c1 in infancy: Associations between maternal caregiving and infant sex. *Infant Mental Health Journal, 40*(4), 513–522.

Cottrell, E. C., & Seckl, J. (2009). Prenatal stress, glucocorticoids and the programming of adult disease. *Frontiers in Behavioral Neuroscience, 3*, 19.

Cox, C. L. (2020). 'Healthcare Heroes': Problems with media focus on heroism from healthcare workers during the COVID-19 pandemic. *Journal of Medical Ethics, 46*(8), 510–513.

Craig, A. D. (2002). How do you feel? Interoception: The sense of The physiological condition of The body. *Nature Reviews. Neuroscience, 3*(8), 655–666.s

Craig, A. D. (2009). How do you feel–now? The anterior insula and human awareness. *Nature Reviews. Neuroscience, 10*(1), 59–70.

Crino, O. L., & Breuner, C. W. (2015). Developmental stress: Evidence for positive phenotypic and fitness effects in birds. *Journal of Ornithology, 156*(1), 389–398.

Crino, O. L., Driscoll, S. C., Ton, R., & Breuner, C. W. (2014). Corticosterone exposure during development improves performance on a novel foraging task in zebra finches. *Animal Behaviour, 91*, 27–32.

Crucianelli, L., Krahé, C., Jenkinson, P. M., & Fotopoulou, A. K. (2018). Interoceptive ingredients of body ownership: Affective touch and cardiac awareness in the rubber hand illusion. *Cortex; a Journal Devoted to the Study of the Nervous System and Behavior, 104*, 180–192.

Crumeyrolle-Arias, M., Jaglin, M., Bruneau, A., Vancassel, S., Cardona, A., Daugé, V., & Rabot, S. (2014). Absence of the gut microbiota enhances anxiety-like behavior and neuroendocrine response to acute stress in rats. *Psychoneuroendocrinology, 42*, 207–217.

Cuffe, J. S., Holland, O., Salomon, C., Rice, G. E., & Perkins, A. V. (2017). Placental derived biomarkers of pregnancy disorders. *Placenta, 54*, 104–110.

Damasio, A. (1994). *Descartes' error: Emotion, reason, and the human brain.* Penguin Group.

Damasio, A. (1999). How the brain creates the mind. *Scientific American, 281*(6), 112–117.

Dancause, K. N., Laplante, D. P., Oremus, C., Fraser, S., Brunet, A., & King, S. (2011). Disaster-related prenatal maternal stress influences birth outcomes: Project Ice Storm. *Early Human Development, 87*(12), 813–820.

Danese, A., Caspi, A., Williams, B., Ambler, A., Sugden, K., Mika, J., & Arseneault, L. (2011). Biological embedding of stress through inflammation processes in childhood. *Molecular Psychiatry, 16*(3), 244–246.

Danese, A., Moffitt, T. E., Pariante, C. M., Ambler, A., Poulton, R., & Caspi, A. (2008). Elevated inflammation levels in depressed adults with a history of childhood maltreatment. *Archives of General Psychiatry, 65*(4), 409–415.

Danese, A., Pariante, C. M., Caspi, A., Taylor, A., & Poulton, R. (2007). Childhood maltreatment predicts adult inflammation in a life-course study. *Proceedings of the National Academy of Sciences, 104*(4), 1319–1324.

Dannlowski, U., Stuhrmann, A., Beutelmann, V., Zwanzger, P., Lenzen, T., Grotegerd, D., & Kugel, H. (2012). Limbic scars: Long-term consequences of childhood maltreatment revealed by functional and structural magnetic resonance imaging. *Biological Psychiatry, 71*(4), 286–293.

Dantzer, R. (2009). Cytokine, sickness behavior, and depression. *Immunology and Allergy Clinics, 29*(2), 247–264.

Darwin, C. (1859). *L'origine della specie.* Torino: Bollati Boringhieri.

Darwin, C. (1872). *Viaggio di un naturalista intorno al mondo.* Unione tipografico-editrice.

Daskalakis, N. P., Bagot, R. C., Parker, K. J., Vinkers, C. H., & de Kloet, E. R. (2013). The three-hit concept of vulnerability and resilience: Toward understanding adaptation to early-life adversity outcome. *Psychoneuroendocrinology, 38*(9), 1858–1873.

Daskalakis, N. P., & Yehuda, R. (2014). Site-specific methylation changes in the glucocorticoid receptor exon 1F promoter in relation to life adversity: Systematic review of contributing factors. *Frontiers in Neuroscience, 8*, 369.

Davis, M., West, K., Bilms, J., Morelen, D., & Suveg, C. (2018). A systematic review of parent–child synchrony: It is more than skin deep. *Developmental Psychobiology, 60*(6), 674–691.

Day, J. J., & Sweatt, J. D. (2011). Cognitive neuroepigenetics: A role for epigenetic mechanisms in learning and memory. *Neurobiology of Learning and Memory, 96*(1), 2–12.

de Bruijn, A. T., van Bakel, H. J., & van Baar, A. L. (2009). Sex differences in the relation between prenatal maternal emotional complaints and child outcome. *Early Human Development, 85*(5), 319–324.

de Jong, T. R., Chauke, M., Harris, B. N., & Saltzman, W. (2009). From here to paternity: Neural correlates of the onset of paternal behavior in California mice (Peromyscus californicus). *Hormones and Behavior, 56*(2), 220–231.

de Kloet, E. R., Oitzl, M. S., & Joëls, M. (1999). Stress and cognition: Are corticosteroids good or bad guys? *Trends in Neurosciences, 22*(10), 422–426.

De Kloet, E. R., Vreugdenhil, E., Oitzl, M. S., & Joëls, M. (1998). Brain corticosteroid receptor balance in health and disease. *Endocrine Reviews, 19*(3), 269–301.

de Rooij, S. R., Painter, R. C., Phillips, D. I., Osmond, C., Michels, R. P., Godsland, I. F., & Roseboom, T. J. (2006). Impaired insulin secretion after prenatal exposure to the Dutch famine. *Diabetes Care, 29*(8), 1897–1901.

de Weerth, C., Fuentes, S., Puylaert, P., & de Vos, W. M. (2013). Intestinal microbiota of infants with colic: Development and specific signatures. *Pediatrics, 131*(2), e550–e558.

De Wolff, M., & van Ijzendoorn, M. H. (1997). Sensitivity and attachment: A meta-analysis on parental antecedents of infant attachment. *Child Development, 68*(4), 571–591.

Dekel, S., Ein-Dor, T., Gordon, K. M., Rosen, J. B., & Bonanno, G. A. (2013). Cortisol and PTSD symptoms among male and female high-exposure 9/11 survivors. *Journal of Traumatic Stress, 26*(5), 621–625.

Del Giudice, M. (2017). Statistical tests of differential susceptibility: Performance, limitations, and improvements. *Development and Psychopathology, 29*(4), 1267–1278.

Del Giudice, M. (2012). Fetal programming by maternal stress: Insights from a conflict perspective. *Psychoneuroendocrinology, 37*(10), 1614–1629.

Del Giudice, M., Ellis, B. J., & Shirtcliff, E. A. (2011). The adaptive calibration model of stress responsivity. *Neuroscience Biobehavioral Reviews, 35*(7), 1562–1592.

Della Longa, L., Filippetti, M. L., Dragovic, D., & Farroni, T. (2020). synchrony of caresses: Does affective touch help infants to detect body-related visual-tactile synchrony? *Frontiers in Psychology, 10*, 2944.

Della Longa, L., Gliga, T., & Farroni, T. (2019). Tune To Touch: Affective Touch enhances learning of face identity in 4-month-old infants. *Developmental Cognitive Neuroscience, 35*, 42–46.

Devlin, A. M., Brain, U., Austin, J., & Oberlander, T. F. (2010). Prenatal exposure to maternal depressed mood and the MTHFR C677T variant affect SLC6A4 methylation in infants at birth. *PLoS One, 5*(8), e12201.

Di Pellegrino, G., Fadiga, L., Fogassi, L., Gallese, V., & Rizzolatti, G. (1992). Understanding motor events: A neurophysiological study. *Experimental Brain Research, 91*(1), 176–180.

Diaz, A., & Bell, M. A. (2012). Frontal EEG asymmetry and fear reactivity in different contexts at 10 months. *Developmental Psychology, 54*(5), 536–545.

Dicorcia, J. A., & Tronick, E. D. (2011). Quotidian resilience: Exploring mechanisms that drive resilience from a perspective of everyday stress and coping. *Neuroscience Biobehavioral Reviews, 35*(7), 1593–1602.

Dipietro, J. A., Costigan, K. A., & Gurewitsch, E. D. (2003). Fetal response to induced maternal stress. *Early Human Development*, *74*(2), 125–138.

Dipietro, J. A., Costigan, K. A., & Voegtline, K. M. (2015). Studies in fetal behavior: Revisited, renewed, and reimagined. *Monographs of the Society for Research in Child Development*, *80*(3), vii.

Ditzen, B., Schaer, M., Gabriel, B., Bodenmann, G., Ehlert, U., & Heinrichs, M. (2009). Intranasal oxytocin increases positive communication and reduces cortisol levels during couple conflict. *Biological Psychiatry*, *65*(9), 728–731.

Dowlati, Y., Herrmann, N., Swardfager, W., Liu, H., Sham, L., Reim, E. K., & Lanctôt, K. L. (2010). A meta-analysis of cytokines in major depression. *Biological Psychiatry*, *67*(5), 446–457.

Drury, S. S., Theall, K., Gleason, M. M., Smyke, A. T., De Vivo, I., Wong, J. Y. Y., & Nelson, C. A. (2012). Telomere length and early severe social deprivation: Linking early adversity and cellular aging. *Molecular Psychiatry*, *17*(7), 719–727.

Dunbar, C. E., High, K. A., Joung, J. K., Kohn, D. B., Ozawa, K., & Sadelain, M. (2018). Gene Therapy comes of age. Science, 359(6372).

Dunn, J., Andrews, C., Nettle, D., & Bateson, M. (2018). Early-life begging effort reduces adult body mass but strengthens behavioural defence of the rate of energy intake in European starlings. *Royal Society Open Science*, *5*(5), 171918.

Dunphy-Doherty, F., O'mahony, S. M., Peterson, V. L., O'sullivan, O., Crispie, F., Cotter, P. D., & Fone, K. C. (2018). Post-weaning social isolation of rats leads to long-term disruption of the gut microbiota-immune-brain axis. *Brain, Behavior, and Immunity*, *68*, 261–273.

Ebisch, S. J., Aureli, T., Bafunno, D., Cardone, D., Romani, G. L., & Merla, A. (2012). Mother and child in synchrony: Thermal facial imprints of autonomic contagion. *Biological Psychology*, *89*(1), 123–129.

Ebralidze, A. K., Rossi, D. J., Tonegawa, S., & Slater, N. T. (1996). Modification of NMDA receptor channels and synaptic transmission by targeted disruption of the NR2C gene. *Journal of Neuroscience*, *16*(16), 5014–5025.

Eccles, J. C. (1964). Presynaptic inhibition in the spinal cord. *Progress in Brain Research*, *12*, 65–91.

Eisen, M. L., Goodman, G. S., Qin, J., Davis, S., & Crayton, J. (2007). Maltreated children's memory: Accuracy, suggestibility, and psychopathology. Developmental Psychology, *43*(6), 1275–1294. https://doi.org/10.1037/0012-1649.43.6.1275

Eisen, M. L., Gabbert, F., Ying, R., & Williams, J. (2017). "I think he had a tattoo on his neck": How co-witness discussions about a Perpetrator's description can affect eyewitness identification decisions. *Journal of Applied Research in Memory and Cognition*, *6*(3), 274–282.

Ellis, B. (2012). Toward an adaptation-based approach to resilience. In J. G. Noll, & I. Shalev (Eds.), (a cura di), *The biology of early stress: Understanding child maltreatment and trauma* (pp. 31–44). Cham: Springer Nature.

Ellis, B. J., Abrams, L. S., Masten, A. S., Sternberg, R. J., Tottenham, N., & Frankenhuis, W. E. (2020). Hidden talents in harsh environments. *Development and Psychopathology*, *1*, 19.

Ellis, B. J., & Boyce, W. T. (2008). Biological sensitivity to context. *Current Directions in Psychological Science*, *17*(3), 183–187.

Ellis, B. J., Boyce, W. T., Belsky, J., Bakermans-Kranenburg, M. J., & van Ijzendoorn, M. H. (2011). Differential susceptibility to the environment: An evolutionary–neurodevelopmental theory. *Development and Psychopathology*, *23*(1), 7–28.

Elman, J. L., Bates, E. A., Johnson, M. H., Karmiloff-Smith, A., et al. (1996). *Rethinking innateness: A connectionist perspective on development*. The MIT Press.

Enoch, M. A., Steer, C. D., Newman, T. K., Gibson, N., & Goldman, D. (2010). Early life stress, MAOA, and gene-environment interactions predict behavioral disinhibition in children. *Genes, Brain and Behavior*, *9*(1), 65–74.

Enthoven, L., Oitzl, M. S., Koning, N., van der Mark, M., & de Kloet, E. R. (2008). Hypothalamic-pituitary-adrenal axis activity of newborn mice rapidly desensitizes to repeated maternal absence but becomes highly responsive to novelty. *Endocrinology, 149*(12), 6366–6377.

Epel, E. S., Blackburn, E. H., Lin, J., Dhabhar, F. S., Adler, N. E., Morrow, J. D., & Cawthon, R. M. (2004). Accelerated telomere shortening in response to life stress. *Proceedings of the National Academy of Sciences, 101*(49), 17312–17315.

Epel, E. S., Lin, J., Dhabhar, F. S., Wolkowitz, O. M., Puterman, E., Karan, L., & Blackburn, E. H. (2010). Dynamics of telomerase activity in response to acute psychological stress. *Brain, Behavior, and Immunity, 24*(4), 531–539.

Epel, E. S., & Prather, A. A. (2018). Stress, telomeres, and psychopathology: Toward a deeper understanding of a triad of early aging. *Annual Review of Clinical Psychology, 14*, 371–397.

Erikson, E. H. (1950). Growth and crises of the "healthy personality." In: Senn, M. J. E. (a cura di), Symposium on the healthy personality (p. 91–146). Josiah Macy, Jr. Foundation.

Evans-Campbell, T. (2008). Historical trauma in American Indian/Native Alaska communities: A multilevel framework for exploring impacts on individuals, families, and communities. *Journal of Interpersonal Violence, 23*(3), 316–338.

Fagiolini, M., Jensen, C. L., & Champagne, F. A. (2009). Epigenetic influences on brain development and plasticity. *Current Opinion in Neurobiology, 19*(2), 207–212.

Fairhurst, M. T., Löken, L., & Grossmann, T. (2014). Physiological and behavioral responses reveal 9-month-old infants' sensitivity to pleasant touch. *Psychological Science, 25*(5), 1124–1131.

Falk, D., & Clarke, R. (2007). Brief communication: New reconstruction of the Taung endocast. *American Journal of Physical Anthropology, 134*(4), 529–534.

Falk, D., Zollikofer, C. P., Morimoto, N., & Ponce de León, M. S. (2012). Metopic suture of Taung (Australopithecus africanus) and its implications for hominin brain evolution. *Proceedings of the National Academy of Sciences of the United States of America, 109*(22), 8467–8470.

Faraone, S. V., Doyle, A. E., Mick, E., & Biederman, J. (2001). Meta-analysis of the association between the 7-repeat allele of the dopamine D4 receptor gene and attention deficit hyperactivity disorder. *American Journal of Psychiatry, 158*(7), 1052–1057.

Farine, D. R., Spencer, K. A., & Boogert, N. J. (2015). Early-life stress triggers juvenile zebra finches to switch social learning strategies. *Current Biology, 25*(16), 2184–2188.

Fatt, P., & Katz, B. (1951). An analysis of the end-plate potential recorded with an intracellular electrode. *The Journal of Physiology, 115*(3), 320–370.

Feldman, R. (2006). From biological rhythms to social rhythms: Physiological precursors of mother-infant synchrony. *Developmental Psychology, 42*(1), 175–188.

Feldman, R. (2007). Parent-infant synchrony and the construction of shared timing; Physiological precursors, developmental outcomes, and risk conditions. *Journal of Child Psychology and Psychiatry, 48*(3-4), 329–354.

Feldman, R. (2012). Parent–infant synchrony: A biobehavioral model of mutual influences in the formation of affiliative bonds. *Monographs of the Society for Research in Child Development, 77*(2), 42–51.

Feldman, R., & Bakermans-Kranenburg, M. J. (2017). Oxytocin: A parenting hormone. *Current Opinion in Psychology, 15*, 13–18.

Feldman, R., Braun, K., & Champagne, F. A. (2019). The neural mechanisms and consequences of paternal caregiving. *Nature Reviews Neuroscience, 20*(4), 205–224.

Feldman, R., & Eidelman, A. I. (2007). Maternal postpartum behavior and the emergence of infant–mother and infant–father synchrony in preterm and full-term infants: The role of neonatal vagal tone. *Developmental Psychobiology, 49*(3), 290–302.

Feldman, R., Gordon, I., Influs, M., Gutbir, T., & Ebstein, R. P. (2013). Parental oxytocin and early caregiving jointly shape children's oxytocin response and social reciprocity. *Neuropsychopharmacology, 38*(7), 1154–1162.

Feldman, R., Greenbaum, C. W., & Yirmiya, N. (1999). Mother–infant affect synchrony as an antecedent of the emergence of self-control. *Developmental Psychology, 35*(1), 223.

Feldman, R., Weller, A., Zagoory-Sharon, O., & Levine, A. (2007). Evidence for a neuroendocrinological foundation of human affiliation: Plasma oxytocin levels across pregnancy and the postpartum period predict mother-infant bonding. *Psychological Science, 18*(11), 965–970.

Feng, X., Wang, L., Yang, S., Qin, D., Wang, J., Li, C., ... & Hu, X. (2011). Maternal separation produces lasting changes in cortisol and behavior in rhesus monkeys. *Proceedings of the National Academy of Sciences, 108*(34), 14312–14317.

Fernandez-Duque, E., Valeggia, C. R., & Mendoza, S. P. (2009). The biology of paternal care in human and nonhuman primates. *Annual Review of Anthropology, 38*, 115–130.

Ferrari, P. F., Gallese, V., Rizzolatti, G., & Fogassi, L. (2003). Mirror neurons responding to the observation of ingestive and communicative mouth actions in the monkey ventral premotor cortex. *The European Journal of Neuroscience, 17*(8), 1703–1714.

Ferretti, P., Pasolli, E., Tett, A., Asnicar, F., Gorfer, V., Fedi, S., & Segata, N. (2018). Mother-to-infant microbial transmission from different body sites shapes the developing infant gut microbiome. *Cell Host Microbe, 24*(1), 133–145.

Field, T. (1994). The effects of mother's physical and emotional unavailability on emotion regulation. *Monographs of the Society for Research in Child Development*, 208–227.

Field, T., & Diego, M. (2008). Cortisol: The culprit prenatal stress variable. *International Journal of Neuroscience, 118*(8), 1181–1205.

Field, D., Draper, E. S., Fenton, A., Papiernik, E., Zeitlin, J., Blondel, B., Van Reempts, P. et al. (2009). Rates of very preterm birth in Europe and neonatal mortality rates. *Archives of Disease in Childhood. Fetal and Neonatal Edition, 94*(4), F253–F256.

Filippa, M., Panza, C., Ferrari, F., Frassoldati, R., Kuhn, P., Balduzzi, S., & D'amico, R. (2017). Systematic review of maternal voice interventions demonstrates increased stability in preterm infants. *Acta Paediatrica, 106*(8), 1220–1229.

Filippetti, M. L., Johnson, M. H., Lloyd-Fox, S., Dragovic, D., & Farroni, T. (2013). Body perception in newborns. *Current Biology, 23*(23), 2413–2416.

Filippetti, M. L., Lloyd-Fox, S., Longo, M. R., Farroni, T., & Johnson, M. H. (2015). Neural mechanisms of body awareness in infants. *Cerebral Cortex, 25*(10), 3779–3787.

Filippi, C. A., Cannon, E. N., Fox, N. A., Thorpe, S. G., Ferrari, P. F., & Woodward, A. L. (2016). Motor system activation predicts goal imitation in 7-month-old infants. *Psychological Science, 27*(5), 675–684.

Finn, S. E. (2009). *Nei panni dei nostri clienti*. Firenze: Giunti OS.

Flacking, R., Ewald, U., Nyqvist, K. H., & Starrin, B. (2006). Trustful bonds: A key to "becoming A mother" and to reciprocal breastfeeding. Stories of mothers of very preterm infants at A neonatal unit. *Social Science Medicine, 62*(1), 70–80.

Fodor, J. A. (1983). The modularity of mind. MIT press.

Fogassi, L., Gallese, V., Di Pellegrino, G., Fadiga, L., Gentilucci, M., Luppino, G., Matelli, M., Pedotti, A., & Rizzolatti, G. (1992). Space coding by premotor cortex. *Experimental Brain Research, 89*(3), 686–690.

Folger, A. T., Ding, L., Ji, H., Yolton, K., Ammerman, R. T., Van Ginkel, J. B., & Bowers, K. (2019). Neonatal NR3C1 methylation and social-emotional development at 6 and 18 months of age. *Frontiers in Behavioral Neuroscience, 13*, 14.

Fonagy, P., Target, M. (2006). The mentalization-focused approach to self pathology. Journal of personality disorders, 20(6), 544–576.

Franke, K., Gaser, C., Roseboom, T. J., Schwab, M., & de Rooij, S. R. (2018). Premature brain aging in humans exposed to maternal nutrient restriction during early gestation. *NeuroImage, 173*, 460–471.

Frankenhuis, W. E., & de Weerth, C. (2013). Does early-life exposure to stress shape or impair cognition? *Current Directions in Psychological Science, 22*(5), 407–412.

Frankenhuis, W. E., Young, E. S., & Ellis, B. J. (2020). The hidden talents approach: Theoretical and methodological challenges. *Trends in Cognitive Sciences, 24*(7), 569–581.

Frazier, W. T., Kandel, E. R., Kupfermann, I., Waziri, R., & Coggeshall, R. E. (1967). Morphological and functional properties of identified neurons in the abdominal ganglion of aplysia californica. *Journal of Neurophysiology, 30*(6), 1288–1351.

Frazzetto, G., Di Lorenzo, G., Carola, V., Proietti, L., Sokolowska, E., Siracusano, A., & Troisi, A. (2007). Early trauma and increased risk for physical aggression during adulthood: The moderating role of MAOA genotype. *PLoS One, 2*(5), e486.

Freud, S. (1892-1899). *Progetto di una psicologia*. Torino: Bollati Boringhieri.

Freud, S. (1914). *Introduzione al narcisismo*. Torino: Bollati Boringhieri.

Freud, S. (1920). *Al di là del principio di piacere*. Torino: Bollati Boringhieri.

Fries, A. B. W., Ziegler, T. E., Kurian, J. R., Jacoris, S., & Pollak, S. D. (2005). Early experience in humans is associated with changes in neuropeptides critical for regulating social behavior. *Proceedings of the National Academy of Sciences, 102*(47), 17237–17240.

Fumagalli, M., Provenzi, L., De Carli, P., Dessimone, F., Sirgiovanni, I., Giorda, R., & Montirosso, R. (2018). From early stress to 12-month development in very preterm infants: Preliminary findings on epigenetic mechanisms and brain growth. *PLoS One, 13*(1), e0190602.

Gallese V. (2007). Embodied simulation: from mirror neuron systems to interpersonal relations. *Novartis Foundation symposium, 278*, 3–221.

Gammill, H. S., Aydelotte, T. M., Guthrie, K. A., Nkwopara, E. C., & Nelson, J. L. (2013). Cellular fetal microchimerism in preeclampsia. *Hypertension, 62*(6), 1062–1067.

Gao, W., Salzwedel, A. P., Carlson, A. L., Xia, K., Azcarate-Peril, M. A., Styner, M. A., & Knickmeyer, R. C. (2019). Gut microbiome and brain functional connectivity in infants-a preliminary study focusing on the amygdala. *Psychopharmacology, 236*(5), 1641–1651.

Gebhardt, C., Mosienko, V., Alenina, N., & Albrecht, D. (2019). Priming of LTP in amygdala and hippocampus by prior paired pulse facilitation paradigm in mice lacking brain serotonin. *Hippocampus, 29*(7), 610–618.

Gergely, G. (2001). The obscure object of desire. Nearly, but clearly not, like me: Contingency preference in normal children versus children with autism. *Bulletin of the Menninger Clinic, 65*(3), 411–426.

Gergely, G., & Watson, J. S. (1996). The social biofeedback Theory of parental affect-mirroring: The development of emotional self-awareness and self-control in infancy. *The International Journal of Psycho-Analysis, 77*(Pt 6), 1181–1212.

Gergely, G., & Watson, J. (1999). Early social-emotional development: Contingency perception and the social biofeedback model. In P. Rochat (Ed.), (a cura di), *Early social cognition: Understanding others in the first months of life*. Hillsdale: Lawrence Erlbaum.

Gilbert, R., Kemp, A., Thoburn, J., Sidebotham, P., Radford, L., Glaser, D., & Macmillan, H. L. (2009). Recognising and responding to child maltreatment. *The Lancet, 373*(9658), 167–180.

Gill, J. M., Saligan, L., Woods, S., & Page, G. (2009). PTSD is associated with an excess of inflammatory immune activities. *Perspectives in Psychiatric Care, 45*(4), 262–277.

Gimsa, U., Tuchscherer, M., & Kanitz, E. (2018). Psychosocial stress and immunity—What can we learn from pig studies? *Frontiers in Behavioral Neuroscience, 12*, 64.

Giorda, R. (2020). Principles of epigenetics and DNA methylation. In Provenzi, L., Montirosso, R. (a cura di), *Developmental human behavioral epigenetics* (pp. 3–26). Academic Press.

Giusti, L., Provenzi, L., & Montirosso, R. (2018). The face-to-face still-face (FFSF) paradigm in clinical settings: Socio-emotional regulation assessment and parental support with infants with neurodevelopmental disabilities. *Frontiers in Psychology, 9*, 789.

Glaser, R., Rabin, B., Chesney, M., Cohen, S., & Natelson, B. (1999). Stress-induced immunomodulation: Implications for infectious diseases? *JAMA*, *281*(24), 2268–2270.

Glasper, E. R., Hyer, M. M., Katakam, J., Harper, R., Ameri, C., & Wolz, T. (2016). Fatherhood contributes to increased hippocampal spine density and anxiety regulation in California mice. *Brain and Behavior*, *6*(1), e00416.

Gluckman, P. D., Wyatt, J. S., Azzopardi, D., Ballard, R., Edwards, A. D., & Ferriero, D. M.COOLCAP STUDY GROUP (2005). Selective head cooling with mild systemic hypothermia after neonatal encephalopathy: Multicentre randomised trial. *The Lancet*, *365*(9460), 663–670.

Glynn, L. M., Davis, E. P., Sandman, C. A., & Goldberg, W. A. (2016). Gestational hormone profiles predict human maternal behavior at 1-year postpartum. *Hormones and Behavior*, *85*, 19–25.

Glynn, L. M., Howland, M. A., & Fox, M. (2018). Maternal programming: Application of a developmental psychopathology perspective. *Development and Psychopathology*, *30*(3), 905–919.

Goerlich, V. C., Nätt, D., Elfwing, M., Macdonald, B., & Jensen, P. (2012). Transgenerational effects of early experience on behavioral, hormonal and gene expression responses to acute stress in the precocial chicken. *Hormones and Behavior*, *61*(5), 711–718.

Goldberg, A. D., Allis, C. D., & Bernstein, E. (2007). Epigenetics: A landscape takes shape. *Cell*, *128*(4), 635–638.

Gomez-Robles, A., Hopkins, W. D., Schapiro, S. J., Sherwood, C. C. (2016). The heritability of chimpanzee and human brain asymmetry. *Proceedings. Biological Sciences*, 283(1845), 20161319.

Gonzalez, A., Jenkins, J. M., Steiner, M., & Fleming, A. S. (2009). The relation between early life adversity, cortisol awakening response and diurnal salivary cortisol levels in postpartum women. *Psychoneuroendocrinology*, *34*(1), 76–86.

Gordon, I., Zagoory-Sharon, O., Leckman, J. F., & Feldman, R. (2010). Oxytocin and the development of parenting in humans. *Biological Psychiatry*, *68*(4), 377–382.

Gould, S. J., & Vrba, E. S. (1982). *Exaptation: Il bricolage dell'evoluzione*. Torino: Bollati Boringhieri.

Govindan, R. M., Behen, M. E., Helder, E., Makki, M. I., & Chugani, H. T. (2010). Altered water diffusivity in cortical association tracts in children with early deprivation identified with tract-based spatial statistics (TBSS). *Cerebral Cortex*, *20*(3), 561–569.

Grant, M. M., White, D., Hadley, J., Hutcheson, N., Shelton, R., Sreenivasan, K., & Deshpande, G. (2014). Early life trauma and directional brain connectivity within major depression. *Human Brain Mapping*, *35*(9), 4815–4826.

Gray, S. A., Jones, C. W., Theall, K. P., Glackin, E., & Drury, S. S. (2017). Thinking across generations: Unique contributions of maternal early life and prenatal stress to infant physiology. *Journal of the American Academy of Child Adolescent Psychiatry*, *56*(11), 922–929.

Grice, S. J., De Haan, M., Halit, H., Johnson, M. H., Csibra, G., Grant, J., & Karmiloff-Smith, A. (2003). ERP abnormalities of illusory contour perception in Williams syndrome. *NeuroReport*, *14*(14), 1773–1777.

Griffiths, B. B., & Hunter, R. G. (2014). Neuroepigenetics of stress. *Neuroscience*, *275*, 420–435.

Grigoriadis, S., VonderPorten, E. H., Mamisashvili, L., Tomlinson, G., Dennis, C. L., Koren, G., & Radford, K. (2013). The impact of maternal depression during pregnancy on perinatal outcomes: A systematic review and meta-analysis. *The Journal of Clinical Psychiatry*, *74*(4), 321–341.

Grossman, P., & Taylor, E. W. (2007). Toward understanding respiratory sinus arrhythmia: Relations to cardiac vagal tone, evolution and biobehavioral functions. *Biological Psychology*, *74*(2), 263–285.

Grumi, S., Saracino, A., Volling, B. L., & Provenzi, L. (2021). A systematic review of human paternal oxytocin: Insights into the methodology and what we know so far. *Developmental Psychobiology*, 63(5), 1330–1344.

Grunau, R. E. (2013). Neonatal pain in very preterm infants: Long-term effects on brain, neurodevelopment and pain reactivity. *Rambam Maimonides Medical Journal*, 4(4).

Gunnar, M. R., Brodersen, L., Nachmias, M., Buss, K., & Rigatuso, J. (1996). Stress reactivity and attachment security. *Developmental Psychobiology*, 29(3), 191–204.

Gunnar, M. R., & Cheatham, C. L. (2003). Brain and behavior interface: Stress and the developing brain. *Infant Mental Health Journal*, 24(3), 195–211.

Gunnar, M. R., & Donzella, B. (2002). Social regulation of the cortisol levels in early human development. *Psychoneuroendocrinology*, 27(1–2), 199–220.

Gunnar, M. R., Gonzalez, C. A., Goodlin, B. L., & Levine, S. (1981). Behavioral and pituitary-adrenal responses during a prolonged separation period in infant rhesus macaques. *Psychoneuroendocrinology*, 6(1), 65–75.

Gunnar, M. R., & Herrera, A. M. (2013). The development of stress reactivity: A neurobiological perspective. In: Zelazo, P.D. (a cura di), Oxford Library of psychology. The Oxford handbook of developmental psychology, vol. 2. Self and other (p. 45–80). Oxford University Press.

Gunnar, M. R., Hertsgaard, L., Larson, M., & Rigatuso, J. (1991). Cortisol and behavioral responses to repeated stressors in the human newborn. *Developmental Psychobiology*, 24(7), 487–505.

Gunnar, M. R., Talge, N. M., & Herrera, A. (2009). Stressor paradigms in developmental studies: What does and does not work to produce mean increases in salivary cortisol. *Psychoneuroendocrinology*, 34(7), 953–967.

Hahn, E., Gottschling, J., & Spinath, F. M. (2013). Current twin studies in Germany: Report on CoSMoS, SOEP, and ChronoS. *Twin Research and Human Genetics*, 16(1), 173–178.

Haig, D. (1993). Genetic conflicts in human pregnancy. *The Quarterly Review of Biology*, 68(4), 495–532.

Haig, D. (1996). Placental hormones, genomic imprinting, and maternal—Fetal communication. *Journal of Evolutionary Biology*, 9(3), 357–380.

Haig, D. (2007). Weismann rules! OK? Epigenetics and the Lamarckian temptation. *Biology Philosophy*, 22(3), 415–428.

Hakeem, A. Y., Sherwood, C. C., Bonar, C. J., Butti, C., Hof, P. R., & Allman, J. M. (2009). Von economo neurons in the elephant brain. *Anatomical Record*, 292(2), 242–248.

Hales, C. N., & Barker, D. J. (1992). Type 2 (non-insulin-dependent) diabetes mellitus: The thrifty phenotype hypothesis. *Diabetologia*, 35(7), 595–601.

Haley, D. W., & Stansbury, K. (2003). Infant stress and parent responsiveness: Regulation of physiology and behavior during still-face and reunion. *Child Development*, 74(5), 1534–1546.

Haley, D. W., Weinberg, J., & Grunau, R. E. (2006). Cortisol, contingency learning, and memory in preterm and full-term infants. *Psychoneuroendocrinology*, 31(1), 108–117.

Halligan, S. L., Herbert, J., Goodyer, I. M., & Murray, L. (2004). Exposure to postnatal depression predicts elevated cortisol in adolescent offspring. *Biological Psychiatry*, 55(4), 376–381.

Halligan, S. L., Murray, L., Martins, C., & Cooper, P. J. (2007). Maternal depression and psychiatric outcomes in adolescent offspring: A 13-year longitudinal study. *Journal of Affective Disorders*, 97(1-3), 145–154.

Ham, J., & Tronick, E. D. (2006). Infant resilience to the stress of the still-face: Infant and maternal psychophysiology are related. *Annals of the New York Academy of Sciences*, 1094(1), 297–302.

Hane, A. A., Lacoursiere, J. N., Mitsuyama, M., Wieman, S., Ludwig, R. J., Kwon, K. Y., & Welch, M. G. (2019). The welch emotional connection screen: Validation of a brief mother–infant relational health screen. *Acta Paediatrica*, 108(4), 615–625.

Hane, A. A., Myers, M. M., Hofer, M. A., Ludwig, R. J., Halperin, M. S., Austin, J., & Welch, M. G. (2015). Family nurture intervention improves the quality of maternal caregiving in the neonatal intensive care unit: Evidence from a randomized controlled trial. *Journal of Developmental Behavioral Pediatrics*, *36*(3), 188–196.

Hantsoo, L., Jasarevic, E., Criniti, S., McGeehan, B., Tanes, C., Sammel, M. D., & Epperson, C. N. (2019). Childhood adversity impact on gut microbiota and inflammatory response to stress during pregnancy. *Brain, Behavior, and Immunity*, *75*, 240–250.

Harris, A., & Seckl, J. (2011). Glucocorticoids, prenatal stress and the programming of disease. *Hormones and Behavior*, *59*(3), 279–289.

Hartwell, K. J., Moran-Santa Maria, M. M., Twal, W. O., Shaftman, S., Desantis, S. M., McRae-Clark, A. L., & Brady, K. T. (2013). Association of elevated cytokines with childhood adversity in a sample of healthy adults. *Journal of Psychiatric Research*, *47*(5), 604–610.

Hebb, D. (1949). *The organization of behaviour*. New York: Wiley.

Heffelfinger, A. K., & Newcomer, J. W. (2001). Glucocorticoid effects on memory function over the human life span. *Development and Psychopathology*, *13*(3), 491–513.

Heijmans, B. T., Tobi, E. W., Lumey, L. H., & Slagboom, P. E. (2009). The epigenome: Archive of the prenatal environment. *Epigenetics*, *4*(8), 526–531.

Heijmans, B. T., Tobi, E. W., Stein, A. D., Putter, H., Blauw, G. J., Susser, E. S., & Lumey, L. H. (2008). Persistent epigenetic differences associated with prenatal exposure to famine in humans. *Proceedings of the National Academy of Sciences*, *105*(44), 17046–17049.

Heim, C. M., Mayberg, H. S., Mletzko, T., Nemeroff, C. B., & Pruessner, J. C. (2013). Decreased cortical representation of genital somatosensory field after childhood sexual abuse. *American Journal of Psychiatry*, *170*(6), 616–623.

Heim, C., Mletzko, T., Purselle, D., Musselman, D. L., & Nemeroff, C. B. (2008). The dexamethasone/corticotropin-releasing factor test in men with major depression: Role of childhood trauma. *Biological Psychiatry*, *63*(4), 398–405.

Heim, C., & Nemeroff, C. B. (1999). The impact of early adverse experiences on brain systems involved in the pathophysiology of anxiety and affective disorders. *Biological Psychiatry*, *46*(11), 1509–1522.

Heim, C., Newport, D. J., Heit, S., Graham, Y. P., Wilcox, M., Bonsall, R., & Nemeroff, C. B. (2000). Pituitary-adrenal and autonomic responses to stress in women after sexual and physical abuse in childhood. *JAMA*, *284*(5), 592–597.

Heim, C., Newport, D. J., Wagner, D., Wilcox, M. M., Miller, A. H., & Nemeroff, C. B. (2002). The role of early adverse experience and adulthood stress in the prediction of neuroendocrine stress reactivity in women: A multiple regression analysis. *Depression and Anxiety*, *15*(3), 117–125.

Heim, C., Owens, M. J., Plotsky, P. M., & Nemeroff, C. B. (1997). Persistent changes in corticotropin-releasing factor systems due to early life stress: Relationship to the pathophysiology of major depression and post-traumatic stress disorder. *Psychopharmacology Bulletin*, *33*(2), 185.

Heim, C., Young, L. J., Newport, D. J., Mletzko, T., Miller, A. H., & Nemeroff, C. B. (2009). Lower CSF oxytocin concentrations in women with a history of childhood abuse. *Molecular Psychiatry*, *14*(10), 954–958.

Hepper, P. G. (1996). Fetal memory: Does it exist? What does it do? *Acta Paediatrica*, *416*, 16–20.

Herculano-Houzel, S., Mota, B., & Lent, R. (2006). Cellular scaling rules for rodent brains. *Proceedings of the National Academy of Sciences of the United States of America*, *103*(32), 12138–12143. https://doi.org/10.1073/pnas.0604911103

Herculano-Houzel, S., Avelino-De-Souza, K., Neves, K., Porfirio, J., Messeder, D., Mattos Feijo, L., Maldonado, J., & Manger, P. R. (2014). The elephant brain in numbers. *Frontiers in Neuroanatomy*, *8*, 46.

Herculano-Houzel, S., Ribeiro, P., Campos, L., Valotta Da Silva, A., Torres, L. B., Catania, K. C., & Kaas, J. H. (2011). Updated neuronal scaling rules for the brains of glires (rodents/lagomorphs). *Brain, Behavior and Evolution, 78*(4), 302–314.

Herman, J. P., Flak, J., & Jankord, R. (2008). Chronic stress plasticity in the hypothalamic paraventricular nucleus. *Progress in Brain Research, 170*, 353–364.

Hertsgaard, L., Gunnar, M., Erickson, M. F., & Nachmias, M. (1995). Adrenocortical responses to the strange situation in infants with disorganized/disoriented attachment relationships. *Child Development, 66*(4), 1100–1106.

Herzenberg, L. A., Bianchi, D. W., Schröder, J., Cann, H. M., & Iverson, G. M. (1979). Fetal cells in the blood of pregnant women: Detection and enrichment by fluorescence-activated cell sorting. *Proceedings of the National Academy of Sciences, 76*(3), 1453–1455.

Heuser, I., Yassouridis, A., & Holsboer, F. (1994). The combined dexamethasone/CRH test: A refined laboratory test for psychiatric disorders. *Journal of Psychiatric Research, 28*(4), 341–356.

Hibel, L. C., Granger, D. A., Blair, C., & Finegood, E. D. Family Life Project Key Investigators (2015). Maternal-child adrenocortical attunement in early childhood: Continuity and change. *Developmental Psychobiology, 57*(1), 83–95.

Hillman, J., & Ventura, M. (1998). *Cento anni di psicoterapia e il mondo va sempre peggio.* Milano: Raffaello Cortina Editore.

Hoban, A. E., Stilling, R. M., Ryan, F. J., Shanahan, F., Dinan, T. G., Claesson, M. J., & Cryan, J. F. (2016). Regulation of prefrontal cortex myelination by the microbiota. *Translational Psychiatry, 6*(4), e774–e774.

Hoekzema, E., Barba-Müller, E., Pozzobon, C., Picado, M., Lucco, F., Garcia-Garcia, D., & Vilarroya, O. (2017). Pregnancy leads to long-lasting changes in human brain structure. *Nature Neuroscience, 20*(2), 287–296.

Hofer, M. A. (2006). Psychobiological roots of early attachment. *Current Directions in Psychological Science, 15*(2), 84–88.

Holmes, C. M., Ghafari, M., Abbas, A., Saravanan, V., & Nemenman, I. (2017). Luria–Delbrück, revisited: The classic experiment does not rule out Lamarckian evolution. *Physical Biology, 14*(5), 055004.

Horn, S. R., Leve, L. D., Levitt, P., & Fisher, P. A. (2019). Childhood adversity, mental health, and oxidative stress: A pilot study. *PLoS One, 14*(4), e0215085.

Howe, M. L., Cicchetti, D., Toth, S. L., & Cerrito, B. M. (2004). True and false memories in maltreated children. *Child Development, 75*(5), 1402–1417.

Howe, M. L., Cicchetti, D., & Toth, S. L. (2006). Children's basic memory processes, stress and maltreatment. *Development and Psychopathology, 18*(3), 759–769.

Howland, M. A., Sandman, C. A., & Glynn, L. M. (2017). Developmental origins of the human hypothalamic-pituitary-adrenal axis. *Expert Review of Endocrinology Metabolism, 12*(5), 321–339.

Hrdy, S. B. (2016). Variable postpartum responsiveness among humans and other primates with "cooperative breeding": A comparative and evolutionary perspective. *Hormones and Behavior, 77*, 272–283.

Hubel, D. H., Wiesel, T. N., & Stryker, M. P. (1977). Orientation columns in macaque monkey visual cortex demonstrated by the 2-deoxyglucose autoradiographic technique. *Nature, 269*(5626), 328–330.

Hucklenbruch-Rother, E., Vohlen, C., Mehdiani, N., Keller, T., Roth, B., Kribs, A., & Mehler, K. (2020). Delivery room skin-to-skin contact in preterm infants affects long-term expression of stress response genes. *Psychoneuroendocrinology, 122*, 104883.

Humphreys, J., Epel, E. S., Cooper, B. A., Lin, J., Blackburn, E. H., & Lee, K. A. (2012). Telomere shortening in formerly abused and never abused women. *Biological Research for Nursing, 14*(2), 115–123.

Hyman, S. E. (2009). How adversity gets under the skin. *Nature Neuroscience, 12*(3), 241–243.

Jabbi, M., Korf, J., Kema, I. P., Hartman, C., van der Pompe, G., Minderaa, R. B., & den Boer, J. A. (2007). Convergent genetic modulation of the endocrine stress response involves polymorphic variations of 5-HTT, COMT and MAOA. *Molecular Psychiatry, 12*(5), 483–490.

Jansen, J., Beijers, R., Riksen-Walraven, M., & de Weerth, C. (2010). Cortisol reactivity in young infants. *Psychoneuroendocrinology, 35*(3), 329–338.

Jardri, R., Houfflin-Debarge, V., Delion, P., Pruvo, J. P., Thomas, P., & Pins, D. (2012). Assessing fetal response to maternal speech using a noninvasive functional brain imaging technique. *International Journal of Developmental Neuroscience, 30*(2), 159–161.

Jean, A. D. L., Stack, D., & Arnold, S. (2014). Investigating maternal touch and infants' self-regulatory behaviours during a modified face-to-face still-face with touch procedure. *Infant and Child Development, 23*(6), 557–574.

Joëls, M., Karst, H., Derijk, R., & de Kloet, E. R. (2008). The coming out of the brain mineralocorticoid receptor. *Trends in Neurosciences, 31*(1), 1–7.

Kananen, L., Surakka, I., Pirkola, S., Suvisaari, J., Lönnqvist, J., Peltonen, L., & Hovatta, I. (2010). Childhood adversities are associated with shorter telomere length at adult age both in individuals with an anxiety disorder and controls. *PLoS One, 5*(5), e10826.

Kandel, E. R. (1976). *Cellular basis of behavior: An introduction to behavioral neurobiology.* W. H. Freeman.

Kandel, E. R. (1998). A new intellectual framework for psychiatry. *American Journal of Psychiatry, 155*(4), 457–469.

Kandel, E. R. (1999). Biology and the future of psychoanalysis: A new intellectual framework for psychiatry revisited. *American Journal of Psychiatry, 156*(4), 505–524.

Kandel, E. R. (2007). *Psichiatria, psicoanalisi e nuova biologia della mente.* Milano: Raffaello Cortina Editore.

Kandel, E. R. (2012). The molecular biology of memory: cAMP, PKA, CRE, CREB-1, CREB-2, and CPEB. *Molecular Brain, 5*(1), 1–12.

Kandel, E. R., Schwartz, J. H., Jessell, T. M. (1992). *Principles of neural science*, 3rd edition. Appleton Lange.

Kantake, M., Yoshitake, H., Ishikawa, H., Araki, Y., & Shimizu, T. (2014). Postnatal epigenetic modification of glucocorticoid receptor gene in preterm infants: A prospective cohort study. *BMJ Open, 4*(7).

Karmiloff-Smith, A. (1992). Learning, development, and conceptual change. Beyond modularity: A developmental perspective on cognitive science. The MIT Press.

Karmiloff-Smith, A. (1998). Development itself is the key to understanding developmental disorders. *Trends in Cognitive Sciences, 2*(10), 389–398.

Karmiloff-Smith, A. (2009). Nativism versus neuroconstructivism: Rethinking the study of developmental disorders. *Developmental Psychology, 45*(1), 56.

Karmiloff-Smith, A. (1997). Crucial differences between developmental cognitive neuroscience and adult neuropsychology. *Developmental Neuropsychology, 13*(4), 513–524.

Kaufman, J. A., Paul, L. K., Manaye, K. F., Granstedt, A. E., Hof, P. R., Hakeem, A. Y., & Allman, J. M. (2008). Selective reduction of von Economo neuron number in agenesis of the corpus callosum. *Acta Neuropathologica, 116*(5), 479–489.

Keenan, K., Gunthorpe, D., & Young, D. (2002). Patterns of cortisol reactivity in African-American neonates from low-income environments. *Developmental Psychobiology, 41*(3), 265–276.

Keenan, D. M., Licinio, J., & Veldhuis, J. D. (2001). A feedback-controlled ensemble model of the stress-responsive hypothalamo-pituitary-adrenal axis. *Proceedings of the National Academy of Sciences, 98*(7), 4028–4033.

Kemp, M. (2003). The Mona Lisa of modern science. *Nature, 421*(6921), 416–420.

Kenis, G., & Maes, M. (2002). Effects of antidepressants on the production of cytokines. *International Journal of Neuropsychopharmacology, 5*(4), 401–412.

Kenkel, W. M., Suboc, G., & Carter, C. S. (2014). Autonomic, behavioral and neuroendocrine correlates of paternal behavior in male prairie voles. *Physiology Behavior, 128*, 252–259.

Keren, M., Feldman, R., Eidelman, A. I., Sirota, L., & Lester, B. (2003). Clinical interview for high-risk parents of premature infants (CLIP) as a predictor of early disruptions in the mother–infant relationship at the nursery. *Infant Mental Health Journal, 24*(2), 93–110.

Kermack, W. O., McKendrick, A. G., & McKinlay, P. L. (1934). Death-rates in great Britain and Sweden: Expression of specific mortality rates as products of two factors, and some consequences thereof. *Epidemiology and Infection, 34*(4), 433–457.

Kim-Cohen, J., Caspi, A., Taylor, A., Williams, B., Newcombe, R., Craig, I. W., & Moffitt, T. E. (2006). MAOA, maltreatment, and gene–environment interaction predicting children's mental health: New evidence and a meta-analysis. *Molecular Psychiatry, 11*(10), 903–913.

King, S., & Laplante, D. P. (2005). The effects of prenatal maternal stress on children's cognitive development: Project ice storm. *Stress, 8*(1), 35–45.

Kinnally, E. L., Lyons, L. A., Abel, K., Mendoza, S., & Capitanio, J. P. (2008). Effects of early experience and genotype on serotonin transporter regulation in infant rhesus macaques. *Genes, Brain, and Behavior, 7*(4), 481–486.

Kinnally, E. L., Martinez, S. J., Chun, K., Capitanio, J. P., & Ceniceros, L. C. (2019). Early social stress promotes inflammation and disease risk in rhesus monkeys. *Scientific Reports, 9*(1), 1–8.

Kinsley, C. H., & Lambert, K. G. (2008). Reproduction-induced neuroplasticity: Natural behavioural and neuronal alterations associated with the production and care of offspring. *Journal of Neuroendocrinology, 20*(4), 515–525.

Kirkpatrick, B., Carter, C. S., Newman, S. W., & Insel, T. R. (1994). Axon-sparing lesions of the medial nucleus of the amygdala decrease affiliative behaviors in the prairie vole (*Microtus ochrogaster*): Behavioral and anatomical specificity. *Behavioral Neuroscience, 108*(3), 501.

Kirschbaum, C., Wolf, O. T., May, M., Wippich, W., & Hellhammer, D. H. (1996). Stress- and treatment-induced elevations of cortisol levels associated with impaired declarative memory in healthy adults. *Life Sciences, 58*(17), 1475–1483.

Kisilevsky, B. S., Hains, S. M., Brown, C. A., Lee, C. T., Cowperthwaite, B., Stutzman, S. S., Swansburg, M. L., Lee, K., Xie, X., Huang, H., Ye, H. H., Zhang, K., & Wang, Z. (2009). Fetal sensitivity to properties of maternal speech and language. *Infant Behavior and Development, 32*(1), 59–71.

Kisilevsky, B. S., Hains, S. M., Lee, K., Xie, X., Huang, H., Ye, H. H., Zhang, K., & Wang, Z. (2003). Effects of experience on fetal voice recognition. *Psychological Science, 14*(3), 220–224.

Kleberg, A., Warren, I., Norman, E., Mörelius, E., Berg, A. C., Mat-Ali, E., & Hellström-Westas, L. (2008). Lower stress responses after newborn individualized developmental care and assessment program care during eye screening examinations for retinopathy of prematurity: A randomized study. *Pediatrics, 121*(5), e1267–e1278.

Klein, M., & Kandel, E. R. (1978). Presynaptic modulation of voltage-dependent Ca2+ current: Mechanism for behavioral sensitization in aplysia californica. *Proceedings of the National Academy of Sciences, 75*(7), 3512–3516.

Klengel, T., & Binder, E. B. (2015). Epigenetics of stress-related psychiatric disorders and gene× environment interactions. *Neuron, 86*(6), 1343–1357.

Kluger, A. N., Siegfried, Z., & Ebstein, R. P. (2002). A meta-analysis of the association between DRD4 polymorphism and novelty seeking. *Molecular Psychiatry, 7*(7), 712–717.

Knickmeyer, R. C., Gouttard, S., Kang, C., Evans, D., Wilber, K., Smith, J. K., Hamer, R. M., Lin, W., Gerig, G., & Gilmore, J. H. (2008). A structural MRI study of human brain development from birth to 2 years. *The Journal of Neuroscience, 28*(47), 12176–12182.

Knuesel, I., Chicha, L., Britschgi, M., Schobel, S. A., Bodmer, M., Hellings, J. A., & Prinssen, E. P. (2014). Maternal immune activation and abnormal brain development across CNS disorders. *Nature Reviews Neurology, 10*(11), 643.

Kochanska, G. (1993). Toward a synthesis of parental socialization and child temperament in early development of conscience. *Child Development, 64*(2), 325–347.

Koenig, J., Kemp, A. H., Feeling, N. R., Thayer, J. F., & Kaess, M. (2016). Resting state vagal tone in borderline personality disorder: A meta-analysis. *Progress in Neuro-Psychopharmacology and Biological Psychiatry, 64*, 18–26.

Konnikova, M. (2016). Why are babies so dumb if humans are so smart? The New Yorker. shorturl.at/firP5

Korraa, A. A., El Nagger, A. A., Mohamed, R. A. E. S., & Helmy, N. M. (2014). Impact of kangaroo mother care on cerebral blood flow of preterm infants. *Italian Journal of Pediatrics, 40*(1), 1–6.

Krzeczkowski, J. E., Van Lieshout, R. J., & Schmidt, L. A. (2020). Transacting brains: Testing an actor–partner model of frontal EEG activity in mother–infant dyads. Development and Psychopathology, 34(3), 969–980.

Kuan, P. F., Waszczuk, M. A., Kotov, R., Marsit, C. J., Guffanti, G., Gonzalez, A., & Luft, B. J. (2017). An epigenome-wide DNA methylation study of PTSD and depression in world trade center responders. *Translational Psychiatry, 7*(6), e1158–e1158.

Kuhlmann, S., Piel, M., & Wolf, O. T. (2005). Impaired memory retrieval after psychosocial stress in healthy young men. *Journal of Neuroscience, 25*(11), 2977–2982.

Labonte, B., Yerko, V., Gross, J., Mechawar, N., Meaney, M. J., Szyf, M., & Turecki, G. (2012). Differential glucocorticoid receptor exon 1B, 1C, and 1H expression and methylation in suicide completers with a history of childhood abuse. *Biological Psychiatry, 72*(1), 41–48.

Ladd, C. O., Huot, R. L., Thrivikraman, K. V., Nemeroff, C. B., Meaney, M. J., & Plotsky, P. M. (2000). Long-term behavioral and neuroendocrine adaptations to adverse early experience. *Progress in brain research, 122*, 81–103.

Laplante, P., Diorio, J., & Meaney, M. J. (2002). Serotonin regulates hippocampal glucocorticoid receptor expression via a 5-HT7 receptor. *Developmental Brain Research, 139*(2), 199–203.

Latva, R., Lehtonen, L., Salmelin, R. K., & Tamminen, T. (2004). Visiting less than every day: A marker for later behavioral problems in Finnish preterm infants. *Archives of Pediatrics Adolescent Medicine, 158*(12), 1153–1157.

Latva, R., Lehtonen, L., Salmelin, R. K., & Tamminen, T. (2007). Visits by the family to the neonatal intensive care unit. *Acta Paediatrica, 96*(2), 215–220.

Lear, J. (2003). La psicoanalisi e i suoi nemici. McGraw-Hill Education.

Leclere, C., Viaux, S., Avril, M., Achard, C., Chetouani, M., Missonnier, S., & Cohen, D. (2014). Why synchrony matters during mother-child interactions: A systematic review. *PLoS One, 9*(12), e113571.

Lederman, S. A., Rauh, V., Weiss, L., Stein, J. L., Hoepner, L. A., Becker, M., & Perera, F. P. (2004). The effects of the World Trade Center event on birth outcomes among term deliveries at three lower Manhattan hospitals. *Environmental Health Perspectives, 112*(17), 1772–1778.

Ledoux, J. E. (2000). Emotion circuits in the brain. *Annual Review of Neuroscience, 23*(1), 155–184.

Lee, A. W., & Brown, R. E. (2002). Medial preoptic lesions disrupt parental behavior in both male and female California mice (peromyscus californicus). *Behavioral Neuroscience, 116*(6), 968.

Lee, A. W., & Brown, R. E. (2007). Comparison of medial preoptic, amygdala, and nucleus accumbens lesions on parental behavior in California mice (peromyscus californicus). *Physiology Behavior, 92*(4), 617–628.

Lee, Y. H., Malakooti, N., & Lotas, M. (2005). A comparison of the light-reduction capacity of commonly used incubator covers. *Neonatal Network*, *24*(2), 37–44.

Lehrner, A., Bierer, L. M., Passarelli, V., Pratchett, L. C., Flory, J. D., Bader, H. N., & Yehuda, R. (2014). Maternal PTSD associates with greater glucocorticoid sensitivity in offspring of holocaust survivors. *Psychoneuroendocrinology*, *40*, 213–220.

Lesch, K. P. (2011). When The serotonin transporter gene meets adversity: The contribution of animal models to understanding epigenetic mechanisms in affective disorders and resilience. Molecular and Functional Models in Neuropsychiatry, 7, 251–280.

Lester, B. M., Conradt, E., Lagasse, L. L., Tronick, E. Z., Padbury, J. F., & Marsit, C. J. (2018). Epigenetic programming by maternal behavior in the human infant. Pediatrics, 142(4).

Lester, B. M., Conradt, E., & Marsit, C. (2016). Introduction to the special section on epigenetics. *Child Development*, *87*(1), 29–37.

Lester, B. M., Tronick, E., Nestler, E., Abel, T., Kosofsky, B., Kuzawa, C. W., & Wood, M. A. (2011). Behavioral epigenetics. *Annals of the New York Academy of Sciences*, *1226*, 14.

Levine, S. (1957). Infantile experience and resistance to physiological stress. *Science*, *126*, 405.

Levine, S. (2005). Developmental determinants of sensitivity and resistance to stress. *Psychoneuroendocrinology*, *30*(10), 939–946.

Levine, S., & Wiener, S. G. (1988). Psychoendocrine aspects of mother-infant relationships in nonhuman primates. *Psychoneuroendocrinology*, *13*(1–2), 143–154.

Levine, A., Zagoory-Sharon, O., Feldman, R., & Weller, A. (2007). Oxytocin during pregnancy and early postpartum: Individual patterns and maternal–fetal attachment. *Peptides*, *28*(6), 1162–1169.

Lindstedt, J., Korja, R., Vilja, S., & Ahlqvist-Björkroth, S. (2020). Fathers' prenatal attachment representations and the quality of father–child interaction in infancy and toddlerhood. Journal of Family Psychology, 35(4), 478–488.

Liu, D., Diorio, J., Day, J. C., Francis, D. D., & Meaney, M. J. (2000). Maternal care, hippocampal synaptogenesis and cognitive development in rats. *Nature Neuroscience*, *3*(8), 799–806.

Liu, D., Diorio, J., Tannenbaum, B., Caldji, C., Francis, D., Freedman, A., & Meaney, M. J. (1997). Maternal care, hippocampal glucocorticoid receptors, and hypothalamic-pituitary-adrenal responses to stress. *Science*, *277*(5332), 1659–1662.

Liu, Y., Ho, R. C. M., & Mak, A. (2012). Interleukin (IL)-6, tumour necrosis factor alpha (TNF-α) and soluble interleukin-2 receptors (sIL-2R) are elevated in patients with major depressive disorder: A meta-analysis and meta-regression. *Journal of Affective Disorders*, *139*(3), 230–239.

Loughman, A., Ponsonby, A. L., O'hely, M., Symeonides, C., Collier, F., Tang, M. L., & Vuillermin, P. (2020). Gut microbiota composition during infancy and subsequent behavioural outcomes. *EBioMedicine*, *52*, 102640.

Lømo T. The discovery of long-term potentiation. Philos Trans R Soc Lond B Biol Sci. 2003 Apr 29;358(1432):617–20. doi: 10.1098/rstb.2002.1226. PMID: 12740104; PMCID: PMC1693150.

Lucas, E. K., & Clem, R. L. (2018). GABAergic interneurons: The orchestra or The conductor in fear learning and memory? *Brain Research Bulletin*, *141*, 13–19.

Lu, B., Kwan, K., Levine, Y. A., Olofsson, P. S., Yang, H., Li, J., & Tracey, K. J. (2014). α7 nicotinic acetylcholine receptor signaling inhibits inflammasome activation by preventing mitochondrial DNA release. *Molecular Medicine*, *20*(1), 350–358.

Lumey, L. H., Stein, A. D., Kahn, H. S., & Romijn, J. A. (2009). Lipid profiles in middle-aged men and women after famine exposure during gestation: The Dutch hunger winter families study. *The American Journal of Clinical Nutrition*, *89*(6), 1737–1743.

Lupien, S. J., King, S., Meaney, M. J., & McEwen, B. S. (2001). Can poverty get under your skin? Basal cortisol levels and cognitive function in children from low and high socioeconomic status. *Development and Psychopathology, 13*(3), 653–676.

Lupien, S. J., McEwen, B. S., Gunnar, M. R., & Heim, C. (2009). Effects of stress throughout the lifespan on the brain, behaviour and cognition. *Nature Reviews Neuroscience, 10*(6), 434–445.

Luria, S. E., & Delbrück, M. (1943). Mutations of bacteria from virus sensitivity to virus resistance. *Genetics, 28*(6), 491.

Lynch, G., Larson, J., Kelso, S., Barrionuevo, G., & Schottler, F. (1983). Intracellular injections of EGTA block induction of hippocampal long-term potentiation. *Nature, 305*(5936), 719–721.

Lyons-Ruth, K. (1998). Implicit relational knowing: Its role in development and psychoanalytic treatment. *Infant Mental Health Journal, 19*(3), 282–289.

Macmillan, H. L., Georgiades, K., Duku, E. K., Shea, A., Steiner, M., Niec, A., & Schmidt, L. A. (2009). Cortisol response to stress in female youths exposed to childhood maltreatment: Results of the youth mood project. *Biological Psychiatry, 66*(1), 62–68.

Macrì, S., & Würbel, H. (2006). Developmental plasticity of HPA and fear responses in rats: A critical review of the maternal mediation hypothesis. *Hormones and Behavior, 50*(5), 667–680.

Maddalena, P. (2013). Long term outcomes of preterm birth: The role of epigenetics. *Newborn and Infant Nursing Reviews, 13*(3), 137–139.

Madden, J. R., & Clutton-Brock, T. H. (2011). Experimental peripheral administration of oxytocin elevates a suite of cooperative behaviours in a wild social mammal. *Proceedings of the Royal Society B: Biological Sciences, 278*(1709), 1189–1194.

Mak, G. K., & Weiss, S. (2010). Paternal recognition of adult offspring mediated by newly generated CNS neurons. *Nature Neuroscience, 13*(6), 753.

Manzotti, A., Cerritelli, F., Esteves, J. E., Lista, G., Lombardi, E., La Rocca, S., Gallace, A., McGlone, F. P., & Walker, S. C. (2019). Dynamic touch reduces physiological arousal in preterm infants: A role for c-tactile afferents? *Developmental Cognitive Neuroscience, 39,* 100703.

Marchesi, J. R., & Ravel, J. (2015). The vocabulary of microbiome research: A proposal. *Microbiome, 3,* 31.

Marlin, B. J., Mitre, M., D'amour, J. A., Chao, M. V., & Froemke, R. C. (2015). Oxytocin enables maternal behaviour by balancing cortical inhibition. *Nature, 520*(7548), 499–504.

Marx, V., & Nagy, E. (2015). Fetal behavioural responses to maternal voice and touch. *PLoS One, 10*(6), e0129118.

Masataka, N. (2001). Why early linguistic milestones are delayed in children with Williams syndrome: Late onset of hand banging as a possible rate–limiting constraint on the emergence of canonical babbling. *Developmental Science, 4*(2), 158–164.

Mascaro, J. S., Rilling, J. K., Tenzin Negi, L., & Raison, C. L. (2013). Compassion meditation enhances empathic accuracy and related neural activity. *Social Cognitive and Affective Neuroscience, 8*(1), 48–55.

Matricardi, S., Agostino, R., Fedeli, C., & Montirosso, R. (2013). Mothers are not fathers: Differences between parents in the reduction of stress levels after a parental intervention in a NICU. *Acta Paediatrica, 102*(1), 8–14.

Matsushita, N., Muroi, Y., Kinoshita, K. I., & Ishii, T. (2015). Comparison of c-fos expression in brain regions involved in maternal behavior of virgin and lactating female mice. *Neuroscience Letters, 590,* 166–171.

Mayford, M., Wang, J., Kandel, E. R., & O'dell, T. J. (1995). CaMKII regulates the frequency-response function of hippocampal synapses for the production of both LTD and LTP. *Cell, 81*(6), 891–904.

Maze, I., & Russo, S. J. (2010). Transcriptional mechanisms underlying addiction-related structural plasticity. *Molecular Interventions, 10*(4), 219.

McCarthy M. Director of top research organization for mental health criticizes DSM for lack of validity *BMJ* 2013; 346 :f2954 doi:10.1136/bmj.f2954

McCarthy, M. I., Abecasis, G. R., Cardon, L. R., Goldstein, D. B., Little, J., Ioannidis, J. P., & Hirschhorn, J. N. (2008). Genome-wide association studies for complex traits: Consensus, uncertainty and challenges. *Nature Reviews Genetics, 9*(5), 356–369.

McCrory, E., De Brito, S. A., & Viding, E. (2010). Research review: The neurobiology and genetics of maltreatment and adversity. *Journal of Child Psychology and Psychiatry, 51*(10), 1079–1095.

McCullough, M. E., Churchland, P. S., & Mendez, A. J. (2013). Problems with measuring peripheral oxytocin: Can the data on oxytocin and human behavior be trusted? *Neuroscience and Biobehavioral Reviews, 37*(8), 1485–1492.

McEwen, B. S. (2000). The neurobiology of stress: From serendipity to clinical relevance. *Brain Research, 886*(1–2), 172–189.

McEwen, B. S. (2006). Protective and damaging effects of stress mediators: central Role of the brain. *Dialogues in Clinical Neuroscience, 8*(4), 367.

McFall-Ngai, M., Hadfield, M. G., Bosch, T. C., Carey, H. V., Domazet-Lošo, T., Douglas, A. E., & Wernegreen, J. J. (2013). Animals in a bacterial world, a new imperative for the life sciences. *Proceedings of the National Academy of Sciences, 110*(9), 3229–3236.

McGlone, F., Wessberg, J., & Olausson, H. (2014). Discriminative and affective touch: Sensing and feeling. *Neuron, 82*(4), 737–755.

McGowan, P. O., Sasaki, A., D'alessio, A. C., Dymov, S., Labonté, B., Szyf, M., & Meaney, M. J. (2009). Epigenetic regulation of the glucocorticoid receptor in human brain associates with childhood abuse. *Nature Neuroscience, 12*(3), 342–348.

McGraw, L. A., & Young, L. J. (2010). The prairie vole: An emerging model organism for understanding the social brain. *Trends in Neurosciences, 33*(2), 103–109.

Meaney, M. J. (2010). Epigenetics and the biological definition of gene× environment interactions. *Child Development, 81*(1), 41–79.

Meaney, D. F., Ross, D. T., Winkelstein, B. A., Brasko, J., Goldstein, D., Bilston, L. B., & Gennarelli, T. A. (1994). Modification of the cortical impact model to produce axonal injury in the rat cerebral cortex. *Journal of Neurotrauma, 11*(5), 599–612.

Meaney, M. J., & Szyf, M. (2005). Environmental programming of stress responses through DNA methylation: Life at the interface between a dynamic environment and a fixed genome. *Dialogues in Clinical Neuroscience, 7*(2), 103–123.

Meimaridou, E., Kowalczyk, J., Guasti, L., Hughes, C. R., Wagner, F., Frommolt, P., & Metherell, L. A. (2012). Mutations in NNT encoding nicotinamide nucleotide transhydrogenase cause familial glucocorticoid deficiency. *Nature Genetics, 44*(7), 740–742.

Meinlschmidt, G., & Heim, C. (2005). Decreased cortisol awakening response after early loss experience. *Psychoneuroendocrinology, 30*(6), 568–576.

Meltzoff, A. N., & Moore, M. K. (1977). Imitation of facial and manual gestures by human neonates. *Science, 198*(4312), 74–78.

Menard, J. L., & Hakvoort, R. M. (2007). Variations of maternal care alter offspring levels of behavioural defensiveness in adulthood: Evidence for a threshold model. *Behavioural Brain Research, 176*(2), 302–313.

Meneses, A., Perez-Garcia, G., Ponce-Lopez, T., Tellez, R., & Castillo, C. (2011). Serotonin transporter and memory. *Neuropharmacology, 61*(3), 355–363.

Mengel-From, J., Thinggaard, M., Dalgård, C., Kyvik, K. O., Christensen, K., & Christiansen, L. (2014). Mitochondrial DNA copy number in peripheral blood cells declines with age and is associated with general health among elderly. *Human Genetics, 133*(9), 1149–1159.

Mercuri, M., Stack, D. M., Trojan, S., Giusti, L., Morandi, F., Mantis, I., & Montirosso, R. (2019). Mothers' and fathers' early tactile contact behaviors during triadic and dyadic parent-infant interactions immediately after birth and at 3-months postpartum: Implications for early care behaviors and intervention. *Infant Behavior and Development, 57,* 101347.

Mervis, C. B., & Velleman, S. L. (2011). Children with Williams syndrome: Language, cognitive, and behavioral characteristics and their implications for intervention. *Perspectives on Language Learning and Education, 18*(3), 98–107.

Mesman, J., van Ijzendoorn, M. H., & Bakermans-Kranenburg, M. J. (2009). The many faces of the still-face paradigm: A review and meta-analysis. *Developmental Review, 29*(2), 120–162.

Messaoudi, M., Violle, N., Bisson, J. F., Desor, D., Javelot, H., & Rougeot, C. (2011). Beneficial psychological effects of a probiotic formulation (*Lactobacillus helveticus* R0052 and *Bifidobacterium longum* R0175) in healthy human volunteers. *Gut Microbes, 2*(4), 256–261.

Midzak, A., & Papadopoulos, V. (2016). Adrenal mitochondria and steroidogenesis: From individual proteins to functional protein assemblies. *Frontiers in Endocrinology, 7,* 106.

Miller, G. (2010). The seductive allure of behavioral epigenetics. *Science, 329,* 24–27.

Mills-Koonce, W. R., Garrett-Peters, P., Barnett, M., Granger, D. A., Blair, C., & Cox, M. J. (2011). Father contributions to cortisol responses in infancy and toddlerhood. *Developmental Psychology, 47*(2), 388–395.

Minolli, M. (2009). *Psicoanalisi della relazione.* Milano: Franco Angeli.

Mittal, C., Griskevicius, V., Simpson, J. A., Sung, S., & Young, E. S. (2015). Cognitive adaptations to stressful environments: When childhood adversity enhances adult executive function. *Journal of Personality and Social Psychology, 109*(4), 604.

Miura, A., Fujiwara, T., Osawa, M., & Anme, T. (2015). Inverse correlation of parental oxytocin levels with autonomy support in toddlers. *Journal of Child and Family Studies, 24*(9), 2620–2625.

Monteiro, D. A., Taylor, E. W., Sartori, M. R., Cruz, A. L., Rantin, F. T., & Leite, C. A. (2018). Cardiorespiratory interactions previously identified as mammalian are present in the primitive lungfish. *Science Advances, 4*(2), eaaq0800.

Montiel-Castro, A. J., González-Cervantes, R. M., Bravo-Ruiseco, G., & Pacheco-López, G. (2013). The microbiota-gut-brain axis: Neurobehavioral correlates, health and sociality. *Frontiers in Integrative Neuroscience, 7,* 70.

Montirosso, R., & McGlone, F. (2020). The body comes first. Embodied reparation and the co-creation of infant bodily-self. *Neuroscience and Biobehavioral Reviews, 113,* 77–87.

Montirosso, R., Piazza, C., Giusti, L., Provenzi, L., Ferrari, P. F., Reni, G., & Borgatti, R. (2019). Exploring the EEG mu rhythm associated with observation and execution of a goal-directed action in 14-month-old preterm infants. *Scientific Reports, 9*(1), 8975.

Montirosso, R., & Provenzi, L. (2015). Implications of epigenetics and stress regulation on research and developmental care of preterm infants. *JOGNN, 44*(2), 174–182.

Montirosso, R., Provenzi, L., Fumagalli, M., Sirgiovanni, I., Giorda, R., Pozzoli, U., & Borgatti, R. (2016a). Serotonin transporter gene (SLC6A4) methylation associates with neonatal intensive care unit stay and 3-month-old temperament in preterm infants. *Child Development, 87*(1), 38–48.

Montirosso, R., Provenzi, L., Giorda, R., Fumagalli, M., Morandi, F., Sirgiovanni, I., & Borgatti, R. (2016b). SLC6A4 promoter region methylation and socio-emotional stress response in very preterm and full-term infants. *Epigenomics, 8*(7), 895–907.

Montirosso, R., Provenzi, L., Tavian, D., Morandi, F., Bonanomi, A., Missaglia, S., & Borgatti, R. (2015). Social stress regulation in 4-month-old infants: Contribution of maternal social engagement and infants' 5-HTTLPR genotype. *Early Human Development, 91*(3), 173–179.

Montirosso, R., Provenzi, L., Tavian, D., Missaglia, S., Raggi, M. E., & Borgatti, R. (2016c). COMTval158met polymorphism is associated with behavioral response and physiologic reactivity to socio-emotional stress in 4-month-old infants. *Infant Behavior and Development*, *45*, 71–82.

Montirosso, R., Provenzi, L., Tronick, E., Morandi, F., Reni, G., & Borgatti, R. (2014). Vagal tone as a biomarker of long-term memory for a stressful social event at 4 months. *Developmental Psychobiology*, *56*(7), 1564–1574.

Montirosso, R., Tronick, E., & Borgatti, R. (2017). Promoting neuroprotective care in neonatal intensive care units and preterm infant development: Insights from the neonatal adequate care for quality of life study. *Child Development Perspectives*, *11*(1), 9–15.

Montirosso, R., Tronick, E., Morandi, F., Ciceri, F., & Borgatti, R. (2013). Four-month-old infants' long-term memory for a stressful social event. *PLoS One*, *8*(12), e82277.

Moody, G., Cannings-John, R., Hood, K., Kemp, A., & Robling, M. (2018). Establishing the international prevalence of self-reported child maltreatment: A systematic review by maltreatment type and gender. *BMC Public Health*, *18*(1), 1–15.

Moore, G. A. (2010). Parent conflict predicts infants' vagal regulation in social interaction. *Development and Psychopathology*, *22*(1), 23.

Moore, G. A., & Calkins, S. D. (2004). Infants' vagal regulation in the still-face paradigm is related to dyadic coordination of mother-infant interaction. *Developmental Psychology*, *40*(6), 1068.

Moore, G. A., Hill-Soderlund, A. L., Propper, C. B., Calkins, S. D., Mills-Koonce, W. R., & Cox, M. J. (2009). Mother–infant vagal regulation in the face-to-face still-face paradigm is moderated by maternal sensitivity. *Child Development*, *80*(1), 209–223.

Mörelius, E., He, H. G., & Shorey, S. (2016). Salivary cortisol reactivity in preterm infants in neonatal intensive care: An integrative review. *International Journal of Environmental Research and Public Health*, *13*(3), 337.

Mörelius, E., Nelson, N., & Gustafsson, P. A. (2007). Salivary cortisol response in mother–infant dyads at high psychosocial risk. *Child: Care, Health and Development*, *33*(2), 128–136.

Mueller, A., Brocke, B., Fries, E., Lesch, K. P., & Kirschbaum, C. (2010). The role of the serotonin transporter polymorphism for the endocrine stress response in newborns. *Psychoneuroendocrinology*, *35*(2), 289–296.

Mueller, E. M., Burgdorf, C., Chavanon, M. L., Schweiger, D., Hennig, J., Wacker, J., & Stemmler, G. (2014). The COMT Val158Met polymorphism regulates the effect of a dopamine antagonist on the feedback-related negativity. *Psychophysiology*, *51*(8), 805–809.

Mulligan, C., D'errico, N., Stees, J., & Hughes, D. (2012). Methylation changes at NR3C1 in newborns associate with maternal prenatal stress exposure and newborn birth weight. *Epigenetics*, *7*(8), 853–857.

Munafò, M. R., Brown, S. M., & Hariri, A. R. (2008). Serotonin transporter (5-HTTLPR) genotype and amygdala activation: A meta-analysis. *Biological Psychiatry*, *63*(9), 852–857.

Murgatroyd, C., Quinn, J. P., Sharp, H. M., Pickles, A., & Hill, J. (2015). Effects of prenatal and postnatal depression, and maternal stroking, at the glucocorticoid receptor gene. *Translational Psychiatry*, *5*(5), e560–e560.

Nakamura, M., Ueno, S., Sano, A., & Tanabe, H. (2000). The human serotonin transporter gene linked polymorphism (5-HTTLPR) shows ten novel allelic variants. *Molecular Psychiatry*, *5*(1), 32–38.

Nanni, V., Uher, R., & Danese, A. (2012). Childhood maltreatment predicts unfavorable course of illness and treatment outcome in depression: A meta-analysis. *American Journal of Psychiatry*, *169*(2), 141–151.

Nazzari, S., Grumi, S., Mambretti, F., Villa, M., Giorda, R., Provenzi, L., & MOM-COPE Study Group Borgatti Renato 1 4 Biasucci Giacomo 5 Decembrino Lidia 6 Giacchero Roberta 7 Magnani Maria Luisa 8 Nacinovich Renata 9 10 Prefumo Federico 11 12 Spinillo Arsenio 13 14 Veggiotti Pierangelo 15 16. (2022). Maternal and infant NR3C1 and SLC6A4 epigenetic signatures of the COVID-19 pandemic lockdown: when timing matters. *Translational psychiatry*, *12*(1), 386.

Nazzari, S., Grumi, S., Biasucci, G., Decembrino, L., Fazzi, E., Giacchero, R., ... & MOM-COPE Study Group. (2023). Maternal pandemic-related stress during pregnancy associates with infants' socio-cognitive development at 12 months: A longitudinal multi-centric study. *Plos one, 18*(4), e0284578.

Nazzi, T., Gopnik, A., & Karmiloff-Smith, A. (2005). Asynchrony in the cognitive and lexical development of young children with Williams syndrome. *Journal of Child Language, 32*(2), 427.

Nazzi, T., Paterson, S., & Karmiloff-Smith, A. (2003). Early word segmentation by infants and toddlers with williams syndrome. *Infancy, 4*(2), 251–271.

Nelson, B. W., Bernstein, R., Allen, N. B., & Laurent, H. K. (2020). The quality of early infant-caregiver relational attachment and longitudinal changes in infant inflammation across 6 months. *Developmental Psychobiology, 62*(5), 674–683.

Nelson, B. W., Wright, D. B., Allen, N. B., & Laurent, H. K. (2020). Maternal stress and social support prospectively predict infant inflammation. *Brain, Behavior, and Immunity, 86*, 14–21.

Nemoda, Z., Massart, R., Suderman, M., Hallett, M., Li, T., Coote, M., & Szyf, M. (2015). Maternal depression is associated with DNA methylation changes in cord blood t lymphocytes and adult hippocampi. *Translational Psychiatry, 5*(4), e545–e545.

Neugebauer, R., Hoek, H. W., & Susser, E. (1999). Prenatal exposure to wartime famine and development of antisocial personality disorder in early adulthood. *JAMA, 282*(5), 455–462.

Ng, S. S., Yue, W. W., Oppermann, U., & Klose, R. J. (2009). Dynamic protein methylation in chromatin biology. *Cellular and Molecular Life Sciences, 66*(3), 407–422.

Nishida, A. H., & Ochman, H. (2018). Rates of gut microbiome divergence in mammals. *Molecular Ecology, 27*(8), 1884–1897.

Nishitani, S., Ikematsu, K., Takamura, T., Honda, S., Yoshiura, K. I., & Shinohara, K. (2017). Genetic variants in oxytocin receptor and arginine-vasopressin receptor 1A are associated with the neural correlates of maternal and paternal affection towards their child. *Hormones and Behavior, 87*, 47–56.

Numan, M. (2012). Maternal behavior: Neural circuits, stimulus Valence, and motivational processes. *Parenting, 12*(2–3), 105–114.

Nussey, D. H., Postma, E., Gienapp, P., & Visser, M. E. (2005). Selection on heritable phenotypic plasticity in a wild bird population. *Science, 310*(5746), 304–306.

O'donnell, K. J., Glover, V., Jenkins, J., Browne, D., Ben-Shlomo, Y., Golding, J., & O'connor, T. G. (2013). Prenatal maternal mood is associated with altered diurnal cortisol in adolescence. *Psychoneuroendocrinology, 38*(9), 1630–1638.

O'donnell, K. J., Jensen, A. B., Freeman, L., Khalife, N., O'connor, T. G., & Glover, V. (2012). Maternal prenatal anxiety and downregulation of placental 11β-HSD2. *Psychoneuroendocrinology, 37*(6), 818–826.

O'donovan, A., Pantell, M. S., Puterman, E., Dhabhar, F. S., Blackburn, E. H., Yaffe, K., & Epel, E. S. (2011). Cumulative inflammatory load is associated with short leukocyte telomere length in the health, aging and body composition study. *PLoS One, 6*(5), e19687.

Oberlander, T. F., Weinberg, J., Papsdorf, M., Grunau, R., Misri, S., & Devlin, A. M. (2008). Prenatal exposure to maternal depression, neonatal methylation of human glucocorticoid receptor gene (NR3C1) and infant cortisol stress responses. *Epigenetics, 3*(2), 97–106.

Oka, T., Hikoso, S., Yamaguchi, O., Taneike, M., Takeda, T., Tamai, T., & Otsu, K. (2012). Mitochondrial DNA that escapes from autophagy causes inflammation and heart failure. *Nature, 485*(7397), 251–255.

Olsson, C. A., Byrnes, G. B., Anney, R. J., Collins, V., Hemphill, S. A., Williamson, R., & Patton, G. C. (2007). COMT Val158Met and 5HTTLPR functional loci interact to predict persistence of anxiety across adolescence: Results from the Victorian adolescent health cohort study. *Genes, Brain and Behavior, 6*(7), 647–652.

Oomen, C. A., Soeters, H., Audureau, N., Vermunt, L., van Hasselt, F. N., Manders, E. M., & Lucassen, P. J. (2011). Early maternal deprivation affects dentate gyrus structure and emotional learning in adult female rats. *Psychopharmacology, 214*(1), 249–260.

Osborne, S., Biaggi, A., Chua, T. E., Du Preez, A., Hazelgrove, K., Nikkheslat, N., & Pariante, C. M. (2018). Antenatal depression programs cortisol stress reactivity in offspring through increased maternal inflammation and cortisol in pregnancy: The psychiatry research and motherhood–depression (PRAM-d) study. *Psychoneuroendocrinology, 98,* 211–221.

Osborne, L. M., Yenokyan, G., Fei, K., Kraus, T., Moran, T., Monk, C., & Sperling, R. (2019). Innate immune activation And depressive And Anxious symptoms across the peripartum: An exploratory study. *Psychoneuroendocrinology, 99,* 80–86.

Ozawa, M., Sasaki, M., & Kanda, K. (2010). Effect of procedure light on the physiological responses of preterm infants. *Japan Journal of Nursing Science, 7*(1), 76–83.

Palma-Gudiel, H., Córdova-Palomera, A., Leza, J. C., & Fañanás, L. (2015). Glucocorticoid receptor gene (NR3C1) methylation processes as mediators of early adversity in stress-related disorders causality: A critical review. *Neuroscience and Biobehavioral Reviews, 55,* 520–535.

Palma-Gudiel, H., & Fañanás, L. (2017). An integrative review of methylation at the serotonin transporter gene and its dialogue with environmental risk factors, psychopathology and 5-HTTLPR. *Neuroscience Biobehavioral Reviews, 72,* 190–209.

Palmieri, P. A., Weathers, F. W., Difede, J., & King, D. W. (2007). Confirmatory factor analysis of the PTSD checklist and the clinician-administered PTSD scale in disaster workers exposed to the world trade center ground zero. *Journal of Abnormal Psychology, 116*(2), 329.

Panagiotidis, J., & Lahav, A. (2010). Simulation of prenatal maternal sounds in NICU incubators: A pilot safety and feasibility study. *The Journal of Maternal-Fetal Neonatal Medicine, 23*(sup3), 106–109.

Paquette, A. G., Lester, B. M., Lesseur, C., Armstrong, D. A., Guerin, D. J., Appleton, A. A., & Marsit, C. J. (2015). Placental epigenetic patterning of glucocorticoid response genes is associated with infant neurodevelopment. *Epigenomics, 7*(5), 767–779.

Parade, S. H., Parent, J., Rabemananjara, K., Seifer, R., Marsit, C. J., Yang, B. Z., & Tyrka, A. R. (2017). Change in FK506 binding protein 5 (FKBP5) methylation over time among preschoolers with adversity. *Development and Psychopathology, 29*(5), 1627.

Parade, S. H., Ridout, K. K., Seifer, R., Armstrong, D. A., Marsit, C. J., McWilliams, M. A., & Tyrka, A. R. (2016). Methylation of the glucocorticoid receptor gene promoter in preschoolers: Links with internalizing behavior problems. *Child Development, 87*(1), 86–97.

Parks, C. G., Miller, D. B., McCanlies, E. C., Cawthon, R. M., Andrew, M. E., Deroo, L. A., & Sandler, D. P. (2009). Telomere length, current perceived stress, and urinary stress hormones in women. *Cancer Epidemiology and Prevention Biomarkers, 18*(2), 551–560.

Partanen, E., Kujala, T., Näätänen, R., Liitola, A., Sambeth, A., & Huotilainen, M. (2013). Learning-induced neural plasticity of speech processing before birth. *Proceedings of the National Academy of Sciences of the United States of America, 110*(37), 15145–15150.

Pärtty, A., Kalliomäki, M., Endo, A., Salminen, S., & Isolauri, E. (2012). Compositional development of bifidobacterium and lactobacillus microbiota is linked with crying and fussing in early infancy. *PloS One, 7*(3), e32495.

Patel, P. D., Lopez, J. F., Lyons, D. M., Burke, S., Wallace, M., & Schatzberg, A. F. (2000). Glucocorticoid and mineralocorticoid receptor mRNA expression in squirrel monkey brain. *Journal of Psychiatric Research, 34*(6), 383–392.

Paukner, A., Simpson, E. A., Ferrari, P. F., Mrozek, T., & Suomi, S. J. (2014). Neonatal imitation predicts how infants engage with faces. *Developmental Science, 17*(6), 833–840.

Pauli-Pott, U., Friedl, S., Hinney, A., & Hebebrand, J. (2009). Serotonin transporter gene polymorphism (5-HTTLPR), environmental conditions, and developing negative emotionality and fear in early childhood. *Journal of Neural Transmission, 116*(4), 503.

Paul, L. K., Schieffer, B., & Brown, W. S. (2004). Social processing deficits in agenesis of the corpus callosum: Narratives from the thematic appreciation test. *Archives of Clinical Neuropsychology, 19*(2), 215–225.

Pawluski, J. L., Lambert, K. G., & Kinsley, C. H. (2016). Neuroplasticity in the maternal hippocampus: Relation to cognition and effects of repeated stress. *Hormones and Behavior*, *77*, 86–97.

Peña, C. J., Monk, C., & Champagne, F. A. (2012). Epigenetic effects of prenatal stress on 11β-hydroxysteroid dehydrogenase-2 in the placenta and fetal brain. *PLoS One*, *7*(6), e39791.

Peng, N. H., Bachman, J., Jenkins, R., Chen, C. H., Chang, Y. C., Chang, Y. S., & Wang, T. M. (2009). Relationships between environmental stressors and stress biobehavioral responses of preterm infants in NICU. *JOGNN*, *23*(4), 363–371.

Peng, H., Zhu, Y., Strachan, E., Fowler, E., Bacus, T., Roy-Byrne, P., & Zhao, J. (2018). Childhood trauma, DNA methylation of stress-related genes, and depression: Findings from two monozygotic twin studies. *Psychosomatic Medicine*, *80*(7), 599.

Perone, S., Gartstein, M. A., & Anderson, A. J. (2020). Dynamics of frontal alpha asymmetry in mother-infant dyads: Insights from the still face paradigm. *Infant Behavior and Development*, *61*, 101500.

Pfefferbaum, B., & North, C. S. (2020). Mental health and the covid-19 pandemic. *New England Journal of Medicine*, *383*(6), 510–512.

Piantadosi, S. T., & Kidd, C. (2016). Extraordinary intelligence and the care of infants. *Proceedings of the National Academy of Sciences of the United States of America*, *113*(25), 6874–6879.

Picard, M., & McEwen, B. S. (2018). Psychological stress and mitochondria: A systematic review. *Psychosomatic Medicine*, *80*(2), 141.

Picard, M., McEwen, B. S., Epel, E. S., & Sandi, C. (2018). An energetic view of stress: Focus on mitochondria. *Frontiers in Neuroendocrinology*, *49*, 72–85.

Picard, M., Shirihai, O. S., Gentil, B. J., & Burelle, Y. (2013). Mitochondrial morphology transitions and functions: Implications for retrograde signaling? *American Journal of Physiology-Regulatory, Integrative and Comparative Physiology*, *304*(6), R393–R406.

Picciolini, O., Porro, M., Meazza, A., Gianni, M. L., Rivoli, C., Lucco, G., & Mosca, F. (2014). Early exposure to maternal voice: Effects on preterm infants development. *Early Human Development*, *90*(6), 287–292.

Pieters, N., Janssen, B. G., Valeri, L., Cox, B., Cuypers, A., Dewitte, H., & Nawrot, T. S. (2015). Molecular responses in the telomere-mitochondrial axis of ageing in the elderly: A candidate gene approach. *Mechanisms of Ageing and Development*, *145*, 51–57.

Pihama, L., Reynolds, P., Smith, C., Reid, J., Smith, L. T., & Nana, R. T. (2014). Positioning historical trauma theory within Aotearoa new zealand. *AlterNative*, *10*(3), 248–262.

Pineda, J. A. (2005). The functional significance of mu rhythms: Translating "seeing" and "hearing" into "doing. *Brain Research: Brain Research Reviews*, *50*(1), 57–68.

Pinkernelle, J., Abraham, A., Seidel, K., & Braun, K. (2009). Paternal deprivation induces dendritic and synaptic changes and hemispheric asymmetry of pyramidal neurons in the somatosensory cortex. *Developmental Neurobiology*, *69*(10), 663–673.

Pinsker, H., Kupfermann, I., Castellucci, V., & Kandel, E. (1970). Habituation and dishabituation of the GM-withdrawal reflex in aplysia. *Science*, *167*(3926), 1740–1742.

Piontelli, A. (2020). *Il culto del feto: Come è cambiata l'immagine della maternità*. Raffaello Cortina Editore: Milano.

Pisoni, C., Provenzi, L., Moncecchi, M., Caporali, C., Naboni, C., Stronati, M., & Orcesi, S. (2021). Early parenting intervention promotes 24-month psychomotor development in preterm children. *Acta Paediatrica*, *110*(1), 101–108.

Plant, D. T., Pariante, C. M., Sharp, D., & Pawlby, S. (2015). Maternal depression during pregnancy and offspring depression in adulthood: Role of child maltreatment. *The British Journal of Psychiatry*, *207*(3), 213–220.

Plant, D. T., Pawlby, S., Sharp, D., Zunszain, P. A., & Pariante, C. M. (2016). Prenatal maternal depression is associated with offspring inflammation at 25 years: A prospective longitudinal cohort study. *Translational Psychiatry*, *6*(11), e936–e936.

Plomin, R. (1983). Introduction: Developmental behavioral genetics. Child Development, 253–259.

Plomin, R. (2013). Child development and molecular genetics: 14 years later. *Child development*, *84*(1), 104–120.

Plomin, R., Defries, J. C., Knopik, V. S., & Neiderhiser, J. M. (2016). Top 10 replicated findings from behavioral genetics. *Perspectives on Psychological Science*, *11*(1), 3–23.

Plotsky, P. M., & Meaney, M. J. (1993). Early, postnatal experience alters hypothalamic corticotropin-releasing factor (CRF) mRNA, median eminence CRF content and stress-induced release in adult rats. *Molecular Brain Research*, *18*(3), 195–200.

Pluess, M., & Belsky, J. (2009). Differential susceptibility to rearing experience: The case of childcare. *Journal of Child Psychology and Psychiatry*, *50*(4), 396–404.

Pluess, M., Velders, F. P., Belsky, J., van Ijzendoorn, M. H., Bakermans-Kranenburg, M. J., Jaddoe, V. W., & Tiemeier, H. (2011). Serotonin transporter polymorphism moderates effects of prenatal maternal anxiety on infant negative emotionality. *Biological Psychiatry*, *69*(6), 520–525.

Pollak, S. D. (2008). Mechanisms linking early experience and the emergence of emotions: Illustrations from the study of maltreated children. *Current Directions in Psychological Science*, *17*(6), 370–375.

Pollitt, R. A., Kaufman, J. S., Rose, K. M., Diez-Roux, A. V., Zeng, D., & Heiss, G. (2007). Early-life and adult socioeconomic status and inflammatory risk markers in adulthood. *European Journal of Epidemiology*, *22*(1), 55–66.

Porges, S. W. (1995). Orienting in a defensive world: Mammalian modifications of our evolutionary heritage. A polyvagal theory. *Psychophysiology*, *32*(4), 301–318.

Porges, S. W. (2003). The polyvagal theory: Phylogenetic contributions to social behavior. *Physiology Behavior*, *79*(3), 503–513.

Porges, S. W. (2011). *The polyvagal theory: Neurophysiological foundations of emotions, attachment, communication, and self-regulation*. WW Norton Company.

Porges, S. W. (2015). Making the world safe for our children: Down-regulating defence and up-regulating social engagement to 'optimise' the human experience. *Children Australia*, *40*(2), 114.

Porges, S. W. (2017). Vagal pathways: Portals to compassion. The Oxford handbook of compassion science, 189–204.

Porges, S. W., & Bohrer, R. E. (1990). The analysis of periodic processes in psychophysiological research. In J. T. Cacioppo L. G. Tassinary (Eds.), *Principles of psychophysiology: Physical, social, and inferential elements* (p. 708–753). Cambridge University Press.

Poutahidis, T., Kearney, S. M., Levkovich, T., Qi, P., Varian, B. J., Lakritz, J. R., & Erdman, S. E. (2013). Microbial symbionts accelerate wound healing via the neuropeptide hormone oxytocin. *PLoS One*, *8*(10), e78898.

Price, L. H., Kao, H. T., Burgers, D. E., Carpenter, L. L., & Tyrka, A. R. (2013). Telomeres and early-life stress: An overview. *Biological Psychiatry*, *73*(1), 15–23.

Prigogine, I., & Stengers, I. (1984). *Order out of chaos: Man's new dialogue with nature*. New York: Bantam Dell Pub Group.

Provenzi, L., & Barello, S. (2015). Behavioral epigenetics of family-centered care in the neonatal intensive care unit. *JAMA Pediatrics*, *169*(7), 697–698.

Provenzi, L., & Barello, S. (2020). The science of the future: Establishing a citizen-scientist collaborative agenda after covid-19. *Frontiers in Public Health*, *8*, 282.

Provenzi, L., Borgatti, R., Menozzi, G., & Montirosso, R. (2015a). A dynamic system analysis of dyadic flexibility and stability across the face-to-face still-face procedure: Application of the state space grid. *Infant Behavior and Development*, *38*, 1–10.

Provenzi, L., Borgatti, R., Montirosso, R. (2016a). The light side of preterm behavioral epigenetics: An epigenetic perspective on caregiver engagement. In: Barello, S., Graffigna, G. (a cura di) *Promoting patient engagement and participation for effective healthcare reform* (pp. 107–127). IGI Global.

Provenzi, L., Borgatti, R., & Montirosso, R. (2017a). Why are prospective longitudinal studies needed in preterm behavioral epigenetic research? *JAMA Pediatrics, 171*(1), 92–92.

Provenzi, L., Brambilla, M., Borgatti, R., & Montirosso, R. (2018a). Methodological challenges in developmental human behavioral epigenetics: Insights into study design. *Frontiers in Behavioral Neuroscience, 12*, 286.

Provenzi, L., Brambilla, M., Scotto Di Minico, G., Montirosso, R., & Borgatti, R. (2020a). Maternal caregiving and DNA methylation in human infants and children: Systematic review. *Genes, Brain and Behavior, 19*(3), e12616.

Provenzi, L., Broso, S., & Montirosso, R. (2018b). Do mothers sound good? A systematic review of the effects of maternal voice exposure on preterm infants' development. *Neuroscience and Biobehavioral Reviews, 88*, 42–50.

Provenzi, L., Casini, E., de Simone, P., Reni, G., Borgatti, R., & Montirosso, R. (2015b). Mother–infant dyadic reparation and individual differences in vagal tone affect 4-month-old infants' social stress regulation. *Journal of Experimental Child Psychology, 140*, 158–170.

Provenzi, L., Cassiano, R. G., Scotto Di Minico, G., Linhares, M., & Montirosso, R. (2017b). Study protocol for the preschooler regulation of emotional stress (PRES) procedure. *Frontiers in Psychology, 8*, 1653.

Provenzi, L., Fumagalli, M., Bernasconi, F., Sirgiovanni, I., Morandi, F., Borgatti, R., & Montirosso, R. (2017c). Very preterm and full-term infants' response to socio-emotional stress: The role of postnatal maternal bonding. *Infancy, 22*(5), 695–712.

Provenzi, L., Fumagalli, M., Giorda, R., Morandi, F., Sirgiovanni, I., Pozzoli, U., & Montirosso, R. (2017d). Maternal sensitivity buffers the association between SLC6A4 methylation and socio-emotional stress response in 3-month-old full term, but not very preterm infants. *Frontiers in Psychiatry, 8*, 171.

Provenzi, L., Fumagalli, M., Scotto Di Minico, G., Giorda, R., Morandi, F., Sirgiovanni, I., & Montirosso, R. (2020b). Pain-related increase in serotonin transporter gene methylation associates with emotional regulation in 4.5-year-old preterm-born children. *Acta Paediatrica, 109*(6), 1166–1174.

Provenzi, L., Fumagalli, M., Sirgiovanni, I., Giorda, R., Pozzoli, U., Morandi, F., & Montirosso, R. (2015c). Pain-related stress during the neonatal intensive care unit stay and SLC6A4 methylation in very preterm infants. *Frontiers in Behavioral Neuroscience, 9*, 99.

Provenzi, L., Giorda, R., Beri, S., & Montirosso, R. (2016b). SLC6A4 methylation as an epigenetic marker of life adversity exposures in humans: A systematic review of literature. *Neuroscience and Biobehavioral Reviews, 71*, 7–20.

Provenzi, L., Giorda, R., Fumagalli, M., Pozzoli, U., Morandi, F., Di Minico, G. S., & Montirosso, R. (2018c). Pain exposure associates with telomere length erosion in very preterm infants. *Psychoneuroendocrinology, 89*, 113–119.

Provenzi, L., Giorda, R., Fumagalli, M., Brambilla, M., Mosca, F., Borgatti, R., & Montirosso, R. (2019a). Telomere length and salivary cortisol stress reactivity in very preterm infants. *Early Human Development, 129*, 1–4.

Provenzi, L., Giusti, L., Fumagalli, M., Tasca, H., Ciceri, F., Menozzi, G., & Montirosso, R. (2016d). Pain-related stress in the neonatal intensive care unit and salivary cortisol reactivity to socio-emotional stress in 3-month-old very preterm infants. *Psychoneuroendocrinology, 72*, 161–165.

Provenzi, L., Giusti, L., Fumagalli, M., Frigerio, S., Morandi, F., Borgatti, R., & Montirosso, R. (2019b). The dual nature of hypothalamic-pituitary-adrenal axis regulation in dyads of very preterm infants and their mothers. *Psychoneuroendocrinology, 100*, 172–179.

Provenzi, L., Giusti, L., & Montirosso, R. (2016c). Do infants exhibit significant cortisol reactivity to the face-to-face still-face paradigm? A narrative review and meta-analysis. *Developmental Review, 42*, 34–55.

Provenzi, L., Grumi, S., & Borgatti, R. (2020c). Alone with the kids: Tele-medicine for children with special healthcare needs during COVID-19 emergency. Frontiers in Psychology, 11.

Provenzi, L., Grumi, S., Giorda, R., Biasucci, G., Bonini, R., Cavallini, A., & Borgatti, R. (2020d). Measuring the outcomes of maternal COVID-19-related prenatal exposure (MOM-COPE): Study protocol for a multicentric longitudinal project. *BMJ Open, 10*(12), e044585.

Provenzi, L., Guida, E., & Montirosso, R. (2018d). Preterm behavioral epigenetics: A systematic review. *Neuroscience Biobehavioral Reviews, 84,* 262–271.

Provenzi, L., Grumi, S., Altieri, L., Bensi, G., Bertazzoli, E., Biasucci, G., ... & MOM-COPE Study Group. (2023). Prenatal maternal stress during the COVID-19 pandemic and infant regulatory capacity at 3 months: A longitudinal study. Development and Psychopathology, *35*(1), 35–43.

Provenzi, L., Mambretti, F., Villa, M., Grumi, S., Citterio, A., Bertazzoli, E., ... & Borgatti, R. (2021). Hidden pandemic: COVID-19-related stress, SLC6A4 methylation, and infants' temperament at 3 months. *Scientific Reports, 11*(1), 15658.

Provenzi, L., & Montirosso, R. (2015). "Epigenethics" in the neonatal intensive care unit: Conveying complexity in health care for preterm children. *JAMA Pediatrics, 169*(7), 617–618.

Provenzi, L., & Montirosso, R. (2020). *Developmental human behavioral epigenetics: Principles, methods, evidence, and future directions.* Academic Press.

Provenzi, L., Rosa, E., Visintin, E., Mascheroni, E., Guida, E., Cavallini, A., & Montirosso, R. (2020e). Understanding the role and function of maternal touch in children with neurodevelopmental disabilities. *Infant Behavior and Development, 58,* 101420.

Provenzi, L., & Santoro, E. (2015). The lived experience of fathers of preterm infants in the neonatal intensive care unit: A systematic review of qualitative studies. *Journal of Clinical Nursing, 24*(13-14), 1784–1794.

Provenzi, L., Scotto Di Minico, G., Giorda, R., & Montirosso, R. (2017e). Telomere length in preterm infants: A promising biomarker of early adversity and care in the neonatal intensive care unit? *Frontiers in Endocrinology, 8,* 295.

Provenzi, L., Scotto Di Minico, G., Giusti, L., Guida, E., & Müller, M. (2018e). Disentangling the dyadic dance: Theoretical, methodological and outcomes systematic review of mother-infant dyadic processes. *Frontiers in Psychology, 9,* 348.

Provenzi, L., & tronick, E. (2020). The power of disconnection during the COVID-19 emergency: From isolation to reparation. *Psychological Trauma: Theory, Research, Practice, and Policy,* 12, S252–S254.

Quadrelli, E., Geangu, E., & Turati, C. (2019). Human action sounds elicit sensorimotor activation early in life. *Cortex, 117,* 323–335.

Quas, J. A., Hong, M., Alkon, A., & Boyce, W. T. (2000). Dissociations between psychobiologic reactivity and emotional expression in children. *Developmental Psychobiology, 37*(3), 153–175.

Radford, L., Corral, S., Bradley, C., & Fisher, H. (2012). Trends in child maltreatment. *The Lancet, 379*(9831), 2048.

Radtke, K. M., Schauer, M., Gunter, H. M., Ruf-Leuschner, M., Sill, J., Meyer, A., & Elbert, T. (2015). Epigenetic modifications of the glucocorticoid receptor gene are associated with the vulnerability to psychopathology in childhood maltreatment. *Translational Psychiatry, 5*(5), e571–e571.

Rakoff, V., Sigal, J. J., & Epstein, N. B. (1966). Children and families of concentration camp survivors. *Canadas Mental Health, 14*(4), 24–26.

Ramsay, D. S., & Lewis, M. (1994). Developmental change in infant cortisol and behavioral response to inoculation. *Child Development, 65*(5), 1491–1502.

Rand, K., & Lahav, A. (2014a). Impact of the NICU environment on language deprivation in preterm infants. *Acta Paediatrica, 103*(3), 243–248.

Rand, K., & Lahav, A. (2014b). Maternal sounds elicit lower heart rate in preterm newborns in the first month of life. *Early Human Development, 90*(10), 679–683.

Ranger, M., & Grunau, R. E. (2014). Early repetitive pain in preterm infants in relation to the developing brain. *Pain Management, 4*(1), 57–67.

Rao, U., Hammen, C., Ortiz, L. R., Chen, L. A., & Poland, R. E. (2008). Effects of early and recent adverse experiences on adrenal response to psychosocial stress in depressed adolescents. *Biological Psychiatry, 64*(6), 521–526.

Rayson, H., Bonaiuto, J. J., Ferrari, P. F., & Murray, L. (2017). Early maternal mirroring predicts infant motor system activation during facial expression observation. *Scientific Reports, 7*(1), 11738.

Ressler, K. J., & Mayberg, H. S. (2007). Targeting abnormal neural circuits in mood and anxiety disorders: From the laboratory to the clinic. *Nature Neuroscience, 10*(9), 1116–1124.

Rhoades, M. W., Reinhart, B. J., Lim, L. P., Burge, C. B., Bartel, B., & Bartel, D. P. (2002). Prediction of plant microRNA targets. *Cell, 110*(4), 513–520.

Richardson, S. S., Daniels, C. R., Gillman, M. W., Golden, J., Kukla, R., Kuzawa, C., & Rich-Edwards, J. (2014). Society: Don't blame the mothers. *Nature News, 512*(7513), 131.

Ridout, K. K., Carpenter, L. L., & Tyrka, A. R. (2016). The cellular sequelae of early stress: Focus on aging and mitochondria. *Neuropsychopharmacology, 41*(1), 388.

Riem, M. M., Bakermans-Kranenburg, M. J., Pieper, S., Tops, M., Boksem, M. A., Vermeiren, R. R., van Ijzendoorn, M. H., & Rombouts, S. A. (2011). Oxytocin modulates amygdala, insula, and inferior frontal gyrus responses to infant crying: A randomized controlled trial. *Biological Psychiatry, 70*(3), 291–297.

Rinn, J. L., & Chang, H. Y. (2012). Genome regulation by long noncoding RNAs. *Annual Review of Biochemistry, 81*, 145–166.

Ritz, T., Enlow, M. B., Schulz, S. M., Kitts, R., Staudenmayer, J., & Wright, R. J. (2012). Respiratory sinus arrhythmia as an index of vagal activity during stress in infants: Respiratory influences and their control. *PLoS One, 7*(12), e52729.

Rizzolatti, G., & Sinigaglia, C. (2005). *So quel che fai: Il cervello che agisce e i neuroni specchio*. Milano: Raffaello Cortina Editore.

Robbins, T. W., & Everitt, B. J. (1999). Drug addiction: Bad habits add up. *Nature, 398*(6728), 567–570.

Robles, T. F., Glaser, R., & Kiecolt-Glaser, J. K. (2005). Out of balance: A new look at chronic stress, depression, and immunity. *Current Directions in Psychological Science, 14*(2), 111–115.

Roberti, E., Giacchero, R., Grumi, S., Biasucci, G., Cuzzani, L., Decembrino, L., ... & Provenzi, L. (2022). Post-partum women's anxiety and parenting stress: Home-visiting protective effect during the COVID-19 pandemic. *Maternal and Child Health Journal, 26*(11), 2308–2317.

Rochat, P., & Striano, T. (2002). Who's in the mirror? Self–other discrimination in specular images by four-and nine-month-Old infants. *Child Development, 73*(1), 35–46.

Rogan, M. T., Stäubli, U. V., & LeDoux, J. E. (1997). AMPA receptor facilitation accelerates fear learning without altering the level of conditioned fear acquired. *Journal of Neuroscience, 17*(15), 5928–5935.

Rogers, F. D., & Bales, K. L. (2019). Mothers, fathers, and others: Neural substrates of parental care. *Trends in Neurosciences, 42*(8), 552–562.

Rogosch, F. A., Dackis, M. N., & Cicchetti, D. (2011). Child maltreatment and allostatic load: Consequences for physical and mental health in children from low-income families. *Development and Psychopathology, 23*(4), 1107.

Rohleder, N. (2014). Stimulation of systemic low-grade inflammation by psychosocial stress. *Psychosomatic Medicine, 76*(3), 181–189.

Roisman, G. I., Newman, D. A., Fraley, R. C., Haltigan, J. D., Groh, A. M., & Haydon, K. C. (2012). Distinguishing differential susceptibility from diathesis–stress: Recommendations for evaluating interaction effects. *Development and Psychopathology, 24*(2), 389–409.

Roseboom, T. J., van der Meulen, J. H., Osmond, C., Barker, D. J., Ravelli, A. C., Schroeder-Tanka, J. M., & Bleker, O. P. (2000). Coronary heart disease after prenatal exposure to the Dutch famine, 1944–45. *Heart, 84*(6), 595–598.

Rosenbeg, K., & Trevathan, W. (1995). Bipedalism and human birth: The obstetrical dilemma revisited. *Evolutionary Anthropology*, *4*(5), 161–168.

Rosen, R. L., Levy-Carrick, N., Reibman, J., Xu, N., Shao, Y., Liu, M., & Galatzer-Levy, I. R. (2017). Elevated c-reactive protein and posttraumatic stress pathology among survivors of the 9/11 world trade center attacks. *Journal of Psychiatric Research*, *89*, 14–21.

Ross, H. E., Cole, C. D., Smith, Y., Neumann, I. D., Landgraf, R., Murphy, A. Z., & Young, L. J. (2009). Characterization of the oxytocin system regulating affiliative behavior in female prairie voles. *Neuroscience*, *162*(4), 892–903.

Roth, M., Neuner, F., & Elbert, T. (2014). Transgenerational consequences of PTSD: Risk factors for the mental health of children whose mothers have been exposed to the Rwandan genocide. *International Journal of Mental Health Systems*, *8*(1), 1–12.

Rottenberg, J., Clift, A., Bolden, S., & Salomon, K. (2007). RSA fluctuation in major depressive disorder. *Psychophysiology*, *44*(3), 450–458.

Sadato, N., Pascual-Leone, A., Grafman, J., Deiber, M. P., Ibanez, V., & Hallett, M. (1998). Neural networks for Braille reading by the blind. *Brain: a journal of neurology*, *121*(7), 1213–1229.

Saito, Y., Fukuhara, R., Aoyama, S., & Toshima, T. (2009). Frontal brain activation in premature infants' response to auditory stimuli in neonatal intensive care unit. *Early Human Development*, *85*(7), 471–474.

Saito, A., & Nakamura, K. (2011). Oxytocin changes primate paternal tolerance to offspring in food transfer. *Journal of Comparative Physiology A*, *197*(4), 329–337.

Sakhai, S. A., Kriegsfeld, L. J., & Francis, D. D. (2011). Maternal programming of sexual attractivity in female Long Evans rats. *Psychoneuroendocrinology*, *36*(8), 1217–1225.

Sanfey, A. G., Hastie, R., Colvin, M. K., & Grafman, J. (2003). Phineas gauged: decision-making and the human prefrontal cortex. *Neuropsychologia*, *41*(9), 1218–1229. https://doi.org/10.1016/s0028-3932(03)00039-3

Santamaria, L., Noreika, V., Georgieva, S., Clackson, K., Wass, S., & Leong, V. (2020). Interpersonal neural network connectivity in mother-infant dyads during emotional communication: A hyperscanning study. Cambridge Neuroscience Seminar, 2019.

Santos, M., Uppal, N., Butti, C., Wicinski, B., Schmeidler, J., Giannakopoulos, P., Heinsen, H., Schmitz, C., & Hof, P. R. (2011). Von economo neurons in autism: A stereologic study of the frontoinsular cortex in children. *Brain Research*, *1380*, 206–217.

Sapolsky, R. M., Romero, L. M., & Munck, A. U. (2000). How do glucocorticoids influence stress responses? Integrating permissive, suppressive, stimulatory, and preparative actions. *Endocrine Reviews*, *21*(1), 55–89.

Sarkar, A., Harty, S., Johnson, K. V. A., Moeller, A. H., Carmody, R. N., Lehto, S. M., & Burnet, P. W. (2020). The role of the microbiome in the neurobiology of social behaviour. *Biological Reviews*, *95*(5), 1131–1166.

Satchell, G. H. (1960). The reflex co-ordination of the heart beat with respiration in the dogfish. *Journal of Experimental Biology*, *37*(4), 719–731.

Savarese, G., Carpinelli, L., Villani, R. A., D'elia, D., & Romei, M. (2020). Data on children involved and the social costs related to the phenomenon of maltreatment and ill-treatment towards children in Italy. *Open Journal of Social Sciences*, *8*(03), 1.

Saxbe, D. E., Golan, O., Ostfeld-Etzion, S., Hirschler-Guttenberg, Y., Zagoory-Sharon, O., & Feldman, R. (2017). HPA axis linkage in parent–child dyads: Effects of parent sex, autism spectrum diagnosis, and dyadic relationship behavior. *Developmental Psychobiology*, *59*(6), 776–786.

Scarr, S., & Weinberg, R. A. (1980). Calling all camps! The war is over. *American Sociological Review*, *45*(5), 859–864.

Schieche, M., & Spangler, G. (2005). Individual differences in biobehavioral organization during problem-solving in toddlers: The influence of maternal behavior, infant–mother attachment, and behavioral inhibition on the attachment-exploration balance. *Developmental Psychobiology*, *46*(4), 293–306.

Schliemann, A. D., & Carraher, D. W. (2002). The evolution of mathematical reasoning: Everyday versus idealized understandings. *Developmental Review, 22*(2), 242–266.

Schneider, U., Bode, F., Schmidt, A., Nowack, S., Rudolph, A., Dölker, E. M., Schlattmann, P., Götz, T., & Hoyer, D. (2018). Developmental milestones of the autonomic nervous system revealed via longitudinal monitoring of fetal heart rate variability. *PLoS One, 13*(7), e0200799.

Sclafani, V., Paukner, A., Suomi, S. J., & Ferrari, P. F. (2015). Imitation promotes affiliation in infant macaques at risk for impaired social behaviors. *Developmental Science, 18*(4), 614–621.

Seeley, W. W., Allman, J. M., Carlin, D. A., Crawford, R. K., Macedo, M. N., Greicius, M. D., Dearmond, S. J., & Miller, B. L. (2007). Divergent social functioning in behavioral variant frontotemporal dementia and Alzheimer disease: Reciprocal networks and neuronal evolution. *Alzheimer Disease and Associated Disorders, 21*(4), S50–S57.

Seeley, W. W., Merkle, F. T., Gaus, S. E., Craig, A. D., Allman, J. M., & Hof, P. R. (2012). Distinctive neurons of the anterior cingulate and frontoinsular cortex: A historical perspective. *Cerebral Cortex, 22*(2), 245–250.

Selye, H. (1936). A syndrome produced by diverse nocuous agents. *Nature, 138*(3479), 32–32.

Selye, H. (1975). Stress and distress. *Comprehensive Therapy, 1*(8), 9–13.

Sender, R., Fuchs, S., & Milo, R. (2016). Are we really vastly outnumbered? Revisiting the ratio of bacterial to host cells in humans. *Cell, 164*(3), 337–340.

Shader, T. M., Gatzke-Kopp, L. M., Crowell, S. E., Jamila Reid, M., Thayer, J. F., Vasey, M. W., & Beauchaine, T. P. (2018). Quantifying respiratory sinus arrhythmia: Effects of misspecifying breathing frequencies across development. *Development and Psychopathology, 30*(1), 351–366.

Shalev, I., Entringer, S., Wadhwa, P. D., Wolkowitz, O. M., Puterman, E., Lin, J., & Epel, E. S. (2013). Stress and telomere biology: A lifespan perspective. *Psychoneuroendocrinology, 38*(9), 1835–1842.

Shammi, M., Bodrud-Doza, M., Islam, A. R. M. T., & Rahman, M. M. (2020). COVID-19 pandemic, socioeconomic crisis and human stress in resource-limited settings: A case from Bangladesh. *Heliyon, 6*(5), e04063.

Sharon, G., Sampson, T. R., Geschwind, D. H., & Mazmanian, S. K. (2016). The central nervous system and the gut microbiome. *Cell, 167*(4), 915–932.

Sharp, H., Hill, J., Hellier, J., & Pickles, A. (2015). Maternal antenatal anxiety, postnatal stroking and emotional problems in children: Outcomes predicted from pre- and postnatal programming hypotheses. *Psychological Medicine, 45*(2), 269–283.

Sheinkopf, S. J., Righi, G., Marsit, C. J., & Lester, B. M. (2016). Methylation of the glucocorticoid receptor (NR3C1) in placenta is associated with infant cry acoustics. *Frontiers in Behavioral Neuroscience, 10*, 100.

Shenk, C. E., Noll, J. G., Putnam, F. W., & Trickett, P. K. (2010). A prospective examination of the role of childhood sexual abuse and physiological asymmetry in the development of psychopathology. *Child Abuse Neglect, 34*(10), 752–761.

Sherrington, R., Brynjolfsson, J., Petursson, H., Potter, M., Dudleston, K., Barraclough, B., & Gurling, H. (1988). Localization of a susceptibility locus for schizophrenia on chromosome 5. *Nature, 336*(6195), 164–167.

Sherwin, E., Bordenstein, S. R., Quinn, J. L., Dinan, T. G., & Cryan, J. F. (2019). Microbiota and the social brain. Science, 366(6465).

Shiio, Y., & Eisenman, R. N. (2003). Histone sumoylation is associated with transcriptional repression. *Proceedings of the National Academy of Sciences, 100*(23), 13225–13230.

Shin KM, Cho SM, Lee SH, Chung YK. A pilot prospective study of the relationship among cognitive factors, shame, and guilt proneness on posttraumatic stress disorder symptoms in female victims of sexual violence. J Korean Med Sci. 2014 Jun;29(6):831–6. doi: 10.3346/jkms.2014.29.6.831. Epub 2014 May 30. PMID: 24932086; PMCID: PMC4055818.

Shinohara, K., & Hata, T. (2014). Post-acquisition hippocampal NMDA receptor blockade sustains retention of spatial reference memory in Morris water maze. *Behavioural Brain Research, 259*, 261–267.

Silver, R. C., Holman, E. A., McIntosh, D. N., Poulin, M., & Gil-Rivas, V. (2002). Nationwide longitudinal study of psychological responses to September 11. *JAMA, 288*(10), 1235–1244.

Simpson, E. A., Fox, N. A., Tramacere, A., & Ferrari, P. F. (2014). Neonatal imitation and an epigenetic account of mirror neuron development. *The Behavioral and Brain Sciences, 37*(2), 220.

Skytthe, A., Christiansen, L., Kyvik, K. O., Bødker, F. L., Hvidberg, L., Petersen, I., ... & Christensen, K. (2013). The Danish Twin Registry: linking surveys, national registers, and biological information. *Twin Research and Human Genetics, 16*(1), 104–111.

Slonim, T. (2014). The polyvagal theory: Neuropsychological foundations of emotions, attachment, communication, self-regulation. *International Journal of Group Psychotherapy, 64*(4), 593–600.

Slopen, N., Kubzansky, L. D., McLaughlin, K. A., & Koenen, K. C. (2013). Childhood adversity and inflammatory processes in youth: A prospective study. *Psychoneuroendocrinology, 38*(2), 188–200.

Slykerman, R. F., Coomarasamy, C., Wickens, K., Thompson, J. M. D., Stanley, T. V., Barthow, C., & Mitchell, E. A. (2019). Exposure to antibiotics in the first 24 months of life and neurocognitive outcomes at 11 years of age. *Psychopharmacology, 236*(5), 1573–1582.

Smith, A. (2016). 'The double helix proves the existence of God': Art and science in dialogue with Salvador Dali's religious imagination. *Approaching Religion, 6*(2), 67–80.

Smith, A. K., Conneely, K. N., Kilaru, V., Mercer, K. B., Weiss, T. E., Bradley, B., & Ressler, K. J. (2011). Differential immune system DNA methylation and cytokine regulation in post-traumatic stress disorder. *American Journal of Medical Genetics Part B: Neuropsychiatric Genetics, 156*(6), 700–708.

Smith, G. C., Gutovich, J., Smyser, C., Pineda, R., Newnham, C., Tjoeng, T. H., & Inder, T. (2011). Neonatal intensive care unit stress is associated with brain development in preterm infants. *Annals of Neurology, 70*(4), 541–549.

Smith, Z. D., & Meissner, A. (2013). DNA methylation: Roles in mammalian development. *Nature Reviews Genetics, 14*(3), 204–220.

Smyke, A. T., Dumitrescu, A., & Zeanah, C. H. (2002). Attachment disturbances in young children. I: The continuum of caretaking casualty. *Journal of the American Academy of Child Adolescent Psychiatry, 41*(8), 972–982.

Spence, S. A., Farrow, T. F., Herford, A. E., Wilkinson, I. D., Zheng, Y., & Woodruff, P. W. (2001). Behavioural and functional anatomical correlates of deception in humans. *Neuroreport, 12*(13), 2849–2853. https://doi.org/10.1097/00001756-200109170-00019

Spitz R, A. (1945). Hospitalism: An inquiry into the genesis of psychiatric conditions in early childhood. *The Psychoanalytic Study of the Child, 1*, 53–74.

Stark, R. I., Garland, M., Daniel, S. S., Tropper, P., & Myers, M. M. (1999). Diurnal rhythms of fetal and maternal heart rate in the baboon. *Early Human Development, 55*(3), 195–209.

Stein, Z., Susser, M., Saenger, G., Marolla, F. (1975). *Famine and human development: The Dutch hunger winter of 1944–1945*. Oxford University Press.

Stenius, F., Theorell, T., Lilja, G., Scheynius, A., Alm, J., & Lindblad, F. (2008). Comparisons between salivary cortisol levels in six-months-olds and their parents. *Psychoneuroendocrinology, 33*(3), 352–359.

Stern, D. (1985). *Interpersonal World Of The Infant: A View From Psychoanalysis And Developmental Psychology*. Basic Books.

Stevens, P., & Eide, M. (1990). The first chapter of children's rights. *American Heritage, 41*(5), 84–91.

Stilling, R. M., Moloney, G. M., Ryan, F. J., Hoban, A. E., Bastiaanssen, T. F., Shanahan, F., & Cryan, J. F. (2018). Social interaction-induced activation of RNA splicing in the amygdala of microbiome-deficient mice. *eLife, 7*, e33070.

Stockley, P., & Hobson, L. (2016). Paternal care and litter size coevolution in mammals. *Proceedings of the Royal Society B: Biological Sciences, 283*(1829), 20160140.

Stoltenberg, S. F., Anderson, C., Nag, P., & Anagnopoulos, C. (2012). Association between the serotonin transporter triallelic genotype and eating problems is moderated by the experience of childhood trauma in women. *International Journal of Eating Disorders, 45*(4), 492–500.

Stolzenberg, D. S., & Champagne, F. A. (2016). Hormonal and non-hormonal bases of maternal behavior: The role of experience and epigenetic mechanisms. *Hormones and Behavior, 77*, 204–210.

Storey, J. D., Robertson, D. A., Beattie, J. E., Reid, I. C., Mitchell, S. N., & Balfour, D. J. (2006). Behavioural and neurochemical responses evoked by repeated exposure to an elevated open platform. *Behavioural Brain Research, 166*(2), 220–229.

Striano, T., Henning, A., & Stahl, D. (2005). Sensitivity to social contingencies between 1 and 3 months of age. *Developmental Science, 8*(6), 509–518.

Strüber, N., Strüber, D., & Roth, G. (2014). Impact of early adversity on glucocorticoid regulation and later mental disorders. *Neuroscience Biobehavioral Reviews, 38*, 17–37.

Sudo, N., Chida, Y., Aiba, Y., Sonoda, J., Oyama, N., Yu, X. N., & Koga, Y. (2004). Postnatal microbial colonization programs the hypothalamic–pituitary–adrenal system for stress response in mice. *The Journal of Physiology, 558*(1), 263–275.

Suomi, S. J. (2006). Risk, resilience, and gene × environment interactions in rhesus monkeys. *Annals of the New York Academy of Sciences, 1094*(1), 52–62.

Susser, E., Neugebauer, R., Hoek, H. W., Brown, A. S., Lin, S., Labovitz, D., & Gorman, J. M. (1996). Schizophrenia after prenatal famine: Further evidence. *Archives of General Psychiatry, 53*(1), 25–31.

Suzuki, M. M., & Bird, A. (2008). DNA methylation landscapes: Provocative insights from epigenomics. *Nature Reviews Genetics, 9*(6), 465–476.

Svob, C., Brown, N. R., Takšić, V., Katulić, K., & Žauhar, V. (2016). Intergenerational transmission of historical memories and social-distance attitudes in post-war second-generation croatians. *Memory and Cognition, 44*(6), 846–855.

Swain, J. E., Ho, S. S., Rosenblum, K. L., Morelen, D., Dayton, C. J., & Muzik, M. (2017). Parent–child intervention decreases stress and increases maternal brain activity and connectivity during own baby-cry: An exploratory study. *Development and Psychopathology, 29*(2), 535.

Szegda, K., Markenson, G., Bertone-Johnson, E. R., & Chasan-Taber, L. (2014). Depression during pregnancy: A risk factor for adverse neonatal outcomes? A critical review of the literature. *The Journal of Maternal-Fetal Neonatal Medicine, 27*(9), 960–967.

Szyf, M., McGowan, P., & Meaney, M. J. (2008). The social environment and the epigenome. *Environmental and Molecular Mutagenesis, 49*(1), 46–60.

Tabbaa, M., Lei, K., Liu, Y., & Wang, Z. (2017). Paternal deprivation affects social behaviors and neurochemical systems in the offspring of socially monogamous prairie voles. *Neuroscience, 343*, 284–297.

Talge, N. M., Neal, C., & Glover, V. (2007). Antenatal maternal stress and long-term effects on child neurodevelopment: How and why? *Journal of Child Psychology and Psychiatry, 48*(3–4), 245–261.

Taylor, E. W. (1992). Nervous control of the heart and cardiorespiratory interactions. *Fish Physiology, 12*, 343–387.

Taylor, S. E., Way, B. M., Welch, W. T., Hilmert, C. J., Lehman, B. J., & Eisenberger, N. I. (2006). Early family environment, current adversity, the serotonin transporter promoter polymorphism, and depressive symptomatology. *Biological Psychiatry, 60*(7), 671–676.

Teicher, M. H., Samson, J. A., Anderson, C. M., & Ohashi, K. (2016). The effects of childhood maltreatment on brain structure, function and connectivity. *Nature Reviews Neuroscience, 17*(10), 652.

Thelen, E., Smith, L. B. (1996). A dynamic systems approach to the development of cognition and action. MIT press.

Thomas, M. P. (1972). Child abuse and neglect: Part 1. Historical overview, legal matrix, and social perspectives. *North Carolina Law Review, 50,* 293–349.

Thomas, R. L., Misra, R., Akkunt, E., Ho, C., Spence, C., & Bremner, A. J. (2018). Sensitivity to auditory-tactile colocation in early infancy. *Developmental science, 21*(4), e12597. https://doi.org/10.1111/desc.12597

Thompson, L. A., Morgan, G., Jurado, K. A., Gunnar, M. R., Bauer, P. J. (2015). A longitudinal study of infant cortisol response during learning events. Monographs of the Society for Research in Child Development, 80(4), 1–143.

Thompson, R. F., & Spencer, W. A. (1966). Habituation: A model phenomenon for the study of neuronal substrates of behavior. *Psychological Review, 73*(1), 16.

Thompson, L. A., & Trevathan, W. R. (2008). Cortisol reactivity, maternal sensitivity, and learning in 3-month-old infants. *Infant Behavior and Development, 31*(1), 92–106.

Thompson, L. A., & Trevathan, W. R. (2009). Cortisol reactivity, maternal sensitivity, and infant preference for mother's familiar face and rhyme in 6-month-old infants. *Journal of Reproductive and Infant Psychology, 27*(2), 143–167.

Thomson, G., Flacking, R., George, K., Feeley, N., Haslund-Thomsen, H., De Coen, K., … Rowe, J. (2020). Parents' experiences of emotional closeness to their infants in the neonatal unit: A meta-ethnography. Early Human Development, 105155.

Tillisch, K., Mayer, E., Gupta, A., Gill, Z., Brazeilles, R., Le Nevé, B., & Labus, J. (2017). Brain structure and response to emotional stimuli as related to gut microbial profiles in healthy women. *Psychosomatic Medicine, 79*(8), 905.

Tobi, E. W., Goeman, J. J., Monajemi, R., Gu, H., Putter, H., Zhang, Y., & Heijmans, B. T. (2014). DNA methylation signatures link prenatal famine exposure to growth and metabolism. *Nature Communications, 5*(1), 1–14.

Tollenaar, M. S., Jansen, J., Beijers, R., Riksen-Walraven, J. M., & de Weerth, C. (2010). Cortisol in the first year of life: Normative values and intra-individual variability. *Early Human Development, 86*(1), 13–16.

Tomoda, A., Polcari, A., Anderson, C. M., & Teicher, M. H. (2012). Reduced visual cortex gray matter volume and thickness in young adults who witnessed domestic violence during childhood. *PLoS One, 7*(12), e52528.

Tomoda, A., Suzuki, H., Rabi, K., Sheu, Y. S., Polcari, A., & Teicher, M. H. (2009). Reduced prefrontal cortical gray matter volume in young adults exposed to harsh corporal punishment. *Neuroimage, 47,* T66–T71.

Toth, S. L., & Cicchetti, D. (2010). The historical origins and developmental pathways of the discipline of developmental psychopathology. *The Israel Journal of Psychiatry and Related Sciences, 47*(2), 95–104.

Tottenham, N., Hare, T. A., Millner, A., Gilhooly, T., Zevin, J. D., & Casey, B. J. (2011). Elevated amygdala response to faces following early deprivation. *Developmental Science, 14*(2), 190–204.

Tozzi, L., Farrell, C., Booij, L., Doolin, K., Nemoda, Z., Szyf, M., & Frodl, T. (2018). Epigenetic changes of FKBP5 as a link connecting genetic and environmental risk factors with structural and functional brain changes in major depression. *Neuropsychopharmacology, 43*(5), 1138–1145.

Trevathan, W. (2015). Primate pelvic anatomy and implications for birth. *Philosophical Transactions of the Royal Society of London. Series B, Biological Sciences, 370*(1663), 20140065.

Trivers, R. L. (1974). Parent-offspring conflict. *Integrative and Comparative Biology, 14*(1), 249–264.

Tronick, E. (2017). The caregiver-infant dyad as a buffer or transducer of resource enhancing or depleting factors that shape psychobiological development. *Australian and New Zealand Journal of Family Therapy*, *38*, 561–572.

Tronick, E. Z. (1989). Emotions and emotional communication in infants. *American Psychologist*, *44*(2), 112.

Tronick, E. Z. (2003). Things still to be done on the still-face effect. *Infancy*, *4*(4), 475–482.

Tronick, E., Als, H., Adamson, L., Wise, S., & Brazelton, T. B. (1978). The infant's response to entrapment between contradictory messages in face-to-face interaction. *Journal of the American Academy of Child Psychiatry*, *17*(1), 1–13.

Tronick, E. Z., & Weinberg, M. K. (1997). Depressed mothers and infants: Failure to form dyadic states of consciousness. In: Murray, L., Cooper, P. J. (a cura di), *Postpartum depression and child development* (p. 54–81). Guilford Press.

Tsankova, N., Renthal, W., Kumar, A., & Nestler, E. J. (2007). Epigenetic regulation in psychiatric disorders. *Nature Reviews Neuroscience*, *8*(5), 355–367.

Turati, C., Simion, F., Milani, I., & Umiltà, C. (2002). Newborns' preference for faces: what is crucial?. *Developmental psychology*, *38*(6), 875–882.

Tuulari, J. J., Scheinin, N. M., Lehtola, S., Merisaari, H., Saunavaara, J., Parkkola, R., Sehlstedt, I., Karlsson, L., Karlsson, H., & Björnsdotter, M. (2019). Neural correlates of gentle skin stroking in early infancy. *Developmental Cognitive Neuroscience*, *35*, 36–41.

Tyrka, A. R., Carpenter, L. L., Kao, H. T., Porton, B., Philip, N. S., Ridout, S. J., & Price, L. H. (2015). Association of telomere length and mitochondrial DNA copy number in a community sample of healthy adults. *Experimental Gerontology*, *66*, 17–20.

Tyrka, A. R., Parade, S. H., Valentine, T. R., Eslinger, N. M., & Seifer, R. (2015). Adversity in preschool-aged children: Effects on salivary interleukin-1β. *Development and Psychopathology*, *27*(2), 567.

Tyrka, A. R., Price, L. H., Kao, H. T., Porton, B., Marsella, S. A., & Carpenter, L. L. (2010). Childhood maltreatment and telomere shortening: Preliminary support for an effect of early stress on cellular aging. *Biological Psychiatry*, *67*(6), 531–534.

Tyrka, A. R., Ridout, K. K., Parade, S. H., Paquette, A., Marsit, C. J., & Seifer, R. (2015). Childhood maltreatment and methylation of FKBP5. *Development and Psychopathology*, *27*(4 Pt 2), 1637.

Tyrka, A. R., Parade, S. H., Valentine, T. R., Eslinger, N. M., & Seifer, R. (2015). Adversity in preschool-aged children: Effects on salivary interleukin-1β. *Development and Psychopathology*, *27*(2), 567–576.

Unternaehrer, E., Bolten, M., Nast, I., Staehli, S., Meyer, A. H., Dempster, E., & Meinlschmidt, G. (2016). Maternal adversities during pregnancy and cord blood oxytocin receptor (OXTR) DNA methylation. *Social Cognitive and Affective Neuroscience*, *11*(9), 1460–1470.

Uvnäs-Moberg, K., Arn, I., & Magnusson, D. (2005). The psychobiology of emotion: The role of the oxytocinergic system. *International Journal of Behavioral Medicine*, *12*(2), 59–65.

Valentino, K., Cicchetti, D., Rogosch, F. A., & Toth, S. L. (2008). True and false recall and dissociation among maltreated children: The role of self-schema. *Development and Psychopathology*, *20*(1), 213–232.

Valkanova, V., Ebmeier, K. P., & Allan, C. L. (2013). CRP, IL-6 and depression: A systematic review and meta-analysis of longitudinal studies. *Journal of Affective Disorders*, *150*(3), 736–744.

van Aken, C., Junger, M., Verhoeven, M., van Aken, M. A. G., & Deković, M. (2007). The interactive effects of temperament and maternal parenting on toddlers' externalizing behaviours. *Infant and Child Development*, *16*(5), 553–572.

van Bakel, H. J., & Riksen-Walraven, J. M. (2002). Parenting and development of one-year-olds: Links with parental, contextual, and child characteristics. *Child Development*, *73*(1), 256–273.

van Bakel, H. J., & Riksen-Walraven, J. M. (2004). Stress reactivity in 15-month-old infants: Links with infant temperament, cognitive competence, and attachment security. *Developmental Psychobiology*, *44*(3), 157–167.

Van Beijsterveldt, C. E., Groen-Blokhuis, M., Hottenga, J. J., Franić, S., Hudziak, J. J., Lamb, D., ... & Boomsma, D. I. (2013). The Young Netherlands Twin Register (YNTR): longitudinal twin and family studies in over 70,000 children. *Twin Research and Human Genetics*, *16*(1), 252–267.

van Dalen, H. P., & Henkens, K. (2020). The COVID-19 pandemic: Lessons for financially fragile and aging societies. *Work, Aging and Retirement*, *6*(4), 229–232.

Van den Bergh, B. R., Van Calster, B., Smits, T., Van Huffel, S., & Lagae, L. (2008). Antenatal maternal anxiety is related to HPA-axis dysregulation and self-reported depressive symptoms in adolescence: A prospective study on the fetal origins of depressed mood. *Neuropsychopharmacology*, *33*(3), 536–545.

van der Vegt, E. J., van der Ende, J., Kirschbaum, C., Verhulst, F. C., & Tiemeier, H. (2009). Early neglect and abuse predict diurnal cortisol patterns in adults: A study of international adoptees. *Psychoneuroendocrinology*, *34*(5), 660–669.

Van Dyck L. I., Morrow E. M. (2017), *Genetic Control of Postnatal Human Brain Growth*, in "Current Opinion in Neurology", 30, 1, pp. 114–24.

van Oers, H. J., de Kloet, E. R., Whelan, T., & Levine, S. (1998). Maternal deprivation effect on the infant's neural stress markers is reversed by tactile stimulation and feeding but not by suppressing corticosterone. *Journal of Neuroscience*, *18*(23), 10171–10179.

Van Puyvelde, M., Gorissen, A. S., Pattyn, N., & McGlone, F. (2019). Does touch matter? The impact of stroking versus non-stroking maternal touch on cardio-respiratory processes in mothers and infants. *Physiology Behavior*, *207*, 55–63.

Veenema, A. H., Koolhaas, J. M., & de Kloet, E. R. (2004). Basal and stress-induced differences in HPA axis, 5-HT responsiveness, and hippocampal cell proliferation in two mouse lines. In 8th Symposium on Catecholamines and Other Neurotransmitters in Stress (pp. 255–265). New York Academy of Sciences.

Venter, J. C., Adams, M. D., Myers, E. W., Li, P. W., Mural, R. J., Sutton, G. G., & Kalush, F. (2001). The sequence of the human genome. *Science*, *291*(5507), 1304–1351.

Vijayendran, M., Beach, S., Plume, J. M., Brody, G., & Philibert, R. (2012). Effects of genotype and child abuse on DNA methylation and gene expression at the serotonin transporter. *Frontiers in Psychiatry*, *3*, 55.

von Bertalanffy, L. (1968). *General system theory: Foundations, development*. New York: George Braziller.

Vythilingam, M., Heim, C., Newport, J., Miller, A. H., Anderson, E., Bronen, R., & Bremner, J. D. (2002). Childhood trauma associated with smaller hippocampal volume in women with major depression. *American Journal of Psychiatry*, *159*(12), 2072–2080.

Wachs, T. D., & Gandour, M. J. (1983). Temperament, environment, and six-month cognitive-intellectual development: A test of the organismic specificity hypothesis. *International Journal of Behavioral Development*, *6*(2), 135–152.

Waddington, C. H. (1942). Canalization of development and the inheritance of acquired characters. *Nature*, *150*(3811), 563–565.

Waddington, C. H. (1961). Molecular biology or ultrastructural biology? *Nature*, *190*(4781), 1124–1125.

Walsh, K., Basu, A., Werner, E., Lee, S., Feng, T., Osborne, L. M., & Monk, C. (2016). Associations among child abuse, depression, and interleukin 6 in pregnant adolescents. *Psychosomatic Medicine*, *78*(8), 920.

Walter, H. (2012). Social cognitive neuroscience of empathy: Concepts, circuits, and genes. *Emotion Review*, *4*(1), 9–17.

Wang, P., Yang, H. P., Tian, S., Wang, L., Wang, S. C., Zhang, F., & Wang, Y. F. (2015). Oxytocin-secreting system: A major part of the neuroendocrine center regulating immunologic activity. *Journal of Neuroimmunology, 289*, 152–161.

Warneken, F., & Tomasello, M. (2006). Altruistic helping in human infants and young chimpanzees. *Science, 311*(5765), 1301–1303.

Wass, S. V., Noreika, V., Georgieva, S., Clackson, K., Brightman, L., Nutbrown, R., & Leong, V. (2018). Parental neural responsivity to infants' visual attention: How mature brains influence immature brains during social interaction. *PLoS Biology, 16*(12), e2006328.

Watkins, S. A. (1990). The Mary Ellen myth: Correcting child welfare history. *Social Work, 35*(6), 500–503.

Weaver, I. C., Cervoni, N., Champagne, F. A., D'alessio, A. C., Sharma, S., Seckl, J. R., & Meaney, M. J. (2004). Epigenetic programming by maternal behavior. *Nature Neuroscience, 7*(8), 847–854.

Webb, A. R., Heller, H. T., Benson, C. B., & Lahav, A. (2015). Mother's voice and heartbeat sounds elicit auditory plasticity in the human brain before full gestation. *Proceedings of the National Academy of Sciences, 112*(10), 3152–3157.

Weinberg, M. K., Tronick, E. Z., Cohn, J. F., & Olson, K. L. (1999). Gender differences in emotional expressivity and self-regulation during early infancy. *Developmental Psychology, 35*(1), 175.

Weinstock, M. (2008). The long-term behavioural consequences of prenatal stress. *Neuroscience Biobehavioral Reviews, 32*(6), 1073–1086.

Weiss, P. (1969). The living system: Determinism stratified. *Alpbach symposium 1968 – Beyond reductionism – New perspectives in the life sciences.* Beacon Press: Boston.

Weisskopf, M. G., Bauer, E. P., & Ledoux, J. E. (1999). L-type voltage-gated calcium channels mediate NMDA-independent associative long-term potentiation at thalamic input synapses to the amygdala. *Journal of Neuroscience, 19*(23), 10512–10519.

Welberg, L. A., & Seckl, J. R. (2001). Prenatal stress, glucocorticoids and the programming of the brain. *Journal of Neuroendocrinology, 13*(2), 113–128.

Welch, M. G., Barone, J. L., Porges, S. W., Hane, A. A., Kwon, K. Y., Ludwig, R. J., & Myers, M. M. (2020). Family nurture intervention in the NICU increases autonomic regulation in mothers and children at 4-5 years of age: Follow-up results from a randomized controlled trial. *PLoS One, 15*(8), e0236930.

Welch, C. D., Check, J., & O'shea, T. M. (2017). Improving care collaboration for NICU patients to decrease length of stay and readmission rate. *BMJ Open Quality, 6*(2), e000130.

Welch, M. G., Halperin, M. S., Austin, J., Stark, R. I., Hofer, M. A., Hane, A. A., & Myers, M. M. (2016). Depression and anxiety symptoms of mothers of preterm infants are decreased at 4 months corrected age with family nurture intervention in the NICU. *Archives of Women's Mental Health, 19*(1), 51–61.

Welch, M. G., Hofer, M. A., Stark, R. I., Andrews, H. F., Austin, J., Glickstein, S. B., & Myers, M. M. (2013). Randomized controlled trial of family nurture intervention in the NICU: Assessments of length of stay, feasibility and safety. *BMC Pediatrics, 13*(1), 1–10.

Wellman, H. M. (1990). *The child's theory of mind.* Cambridge: MIT Press.

West-Eberhard, M. J. (2003). Developmental plasticity and evolution. Oxford University Press.

Westrup, B. (2007). Newborn individualized developmental care and assessment program (NIDCAP)—family-centered developmentally supportive care. *Early Human Development, 83*(7), 443–449.

Wheeler, E. A. (1910). *The story of Mary Ellen.* The American Humane Association, 9725 Denver.

Widom, C. S., & Brzustowicz, L. M. (2006). MAOA and the "cycle of violence:" Childhood abuse and neglect, MAOA genotype, and risk for violent and antisocial behavior. *Biological Psychiatry, 60*(7), 684–689.

Wiesel, T. N., & Hubel, D. H. (1963). Single-cell responses in striate cortex of kittens deprived of vision in one eye. *Journal of Neurophysiology, 26*, 1003–1017.

Wikoff, W. R., Anfora, A. T., Liu, J., Schultz, P. G., Lesley, S. A., Peters, E. C., & Siuzdak, G. (2009). Metabolomics analysis reveals large effects of gut microflora on mammalian blood metabolites. *Proceedings of the National Academy of Sciences, 106*(10), 3698–3703.

Winberg, J. (2005). Mother and newborn baby: Mutual regulation of physiology and behavior–a selective review. *Developmental Psychobiology, 47*(3), 217–229.

Winslow, J. T., Noble, P. L., Lyons, C. K., Sterk, S. M., & Insel, T. R. (2003). Rearing effects on cerebrospinal fluid oxytocin concentration and social buffering in rhesus monkeys. *Neuropsychopharmacology, 28*(5), 910–918.

Wirleitner, B., Neurauter, G., Schrocksnadel, K., Frick, B., & Fuchs, D. (2003). Interferon-γ-induced conversion of tryptophan: Immunologic and neuropsychiatric aspects. *Current Medicinal Chemistry, 10*(16), 1581–1591.

Wolff, G. L., Kodell, R. L., Moore, S. R., & Cooney, C. A. (1998). Maternal epigenetics and methyl supplements affect agouti gene expression in Avy/a mice. *The FASEB Journal, 12*(11), 949–957.

Wolff, B. C., Wadsworth, M. E., Wilhelm, F. H., & Mauss, I. B. (2012). Children's vagal regulatory capacity predicts attenuated sympathetic stress reactivity in a socially supportive context: Evidence for a protective effect of the vagal system. *Development and Psychopathology, 24*(2), 677–689.

Wu, R., Song, Z., Wang, S., Shui, L., Tai, F., Qiao, X., & He, F. (2014). Early paternal deprivation alters levels of hippocampal brain-derived neurotrophic factor and glucocorticoid receptor and serum corticosterone and adrenocorticotropin in a sex-specific way in socially monogamous mandarin voles. *Neuroendocrinology, 100*(2–3), 119–128.

Yang, C., Fujita, Y., Ren, Q., Ma, M., Dong, C., & Hashimoto, K. (2017). Bifidobacterium in the gut microbiota confer resilience to chronic social defeat stress in mice. *Scientific Reports, 7*(1), 1–7.

Yaroslavsky, I., Bylsma, L. M., Rottenberg, J., & Kovacs, M. (2013). Combinations of resting RSA and RSA reactivity impact maladaptive mood repair and depression symptoms. *Biological Psychology, 94*(2), 272–281.

Yehuda, R., Bell, A., Bierer, L. M., & Schmeidler, J. (2008). Maternal, not paternal, PTSD is related to increased risk for PTSD in offspring of Holocaust survivors. *Journal of Psychiatric Research, 42*(13), 1104–1111.

Yehuda, R., & Bierer, L. M. (2009). The relevance of epigenetics to PTSD: Implications for the DSM-V. *Journal of Traumatic Stress, 22*(5), 427–434.

Yehuda, R., Cai, G., Golier, J. A., Sarapas, C., Galea, S., Ising, M., & Buxbaum, J. D. (2009). Gene Expression patterns associated with posttraumatic stress disorder following exposure to the World Trade Center attacks. *Biological Psychiatry, 66*(7), 708–711.

Yehuda, R., Daskalakis, N. P., Bierer, L. M., Bader, H. N., Klengel, T., Holsboer, F., & Binder, E. B. (2016). Holocaust exposure induced intergenerational effects on FKBP5 methylation. *Biological Psychiatry, 80*(5), 372–380.

Yehuda, R., Daskalakis, N. P., Desarnaud, F., Makotkine, I., Lehrner, A., Koch, E., & Bierer, L. M. (2013). Epigenetic biomarkers as predictors and correlates of symptom improvement following psychotherapy in combat veterans with PTSD. *Frontiers in Psychiatry, 4*, 118.

Yehuda, R., Engel, S. M., Brand, S. R., Seckl, J., Marcus, S. M., & Berkowitz, G. S. (2005). Transgenerational effects of posttraumatic stress disorder in babies of mothers exposed to the World Trade Center attacks during pregnancy. *The Journal of Clinical Endocrinology Metabolism, 90*(7), 4115–4118.

Yehuda, R., Halligan, S. L., & Bierer, L. M. (2001). Relationship of parental trauma exposure and PTSD to PTSD, depressive and anxiety disorders in offspring. *Journal of psychiatric research, 35*(5), 261–270.

Yehuda, R., Daskalakis, N. P., Lehrner, A., Desarnaud, F., Bader, H. N., Makotkine, I., ... & Meaney, M. J. (2014). Influences of maternal and paternal PTSD on epigenetic regulation of the glucocorticoid receptor gene in Holocaust survivor offspring. *American Journal of Psychiatry, 171*(8), 872–880.

Yehuda, R., Bierer, L. M., Schmeidler, J., Aferiat, D. H., Breslau, I., & Dolan, S. (2000). Low cortisol and risk for PTSD in adult offspring of holocaust survivors. *American Journal of Psychiatry, 157*(8), 1252–1259.

Yehuda, R., Halligan, S. L., & Bierer, L. M. (2002). Cortisol levels in adult offspring of holocaust survivors: Relation to PTSD symptom severity in the parent and child. *Psychoneuroendocrinology, 27*(1–2), 171–180.

Yehuda, R., Halligan, S. L., Golier, J. A., Grossman, R., & Bierer, L. M. (2004). Effects of trauma exposure on the cortisol response to dexamethasone administration in PTSD and major depressive disorder. *Psychoneuroendocrinology, 29*(3), 389–404.

Yehuda, R., Kahana, B., Schmeidler, J., Southwick, S. M., Wilson, S., & Giller, E. L. (1995). Impact of cumulative lifetime trauma and recent stress on current posttraumatic stress disorder symptoms in holocaust survivors. *The American Journal of Psychiatry, 152*(12), 1815–1818.

Yehuda, R., Morris, A., Labinsky, E., Zemelman, S., & Schmeidler, J. (2007). Ten-year follow-up study of cortisol levels in aging holocaust survivors with and without PTSD. *Journal of Traumatic Stress, 20*(5), 757–761.

Yehuda, R., Teicher, M. H., Seckl, J. R., Grossman, R. A., Morris, A., & Bierer, L. M. (2007). Parental posttraumatic stress disorder as a vulnerability factor for low cortisol trait in offspring of holocaust survivors. *Archives of General Psychiatry, 64*(9), 1040–1048.

Yu, P., An, S., Tai, F., Zhang, X., He, F., Wang, J., & Wu, R. (2012). The effects of neonatal paternal deprivation on pair bonding, NAcc dopamine receptor mRNA expression and serum corticosterone in mandarin voles. *Hormones and Behavior, 61*(5), 669–677.

Zeanah, C. H. (2000). Disturbances of attachment in young children adopted from institutions. *Journal of Developmental and Behavioral Pediatrics, 21*(3), 230–236.

Zhou, M., Bakker, E. H., Velzing, E. H., Berger, S., Oitzl, M., Joëls, M., & Krugers, H. J. (2010). Both mineralocorticoid and glucocorticoid receptors regulate emotional memory in mice. *Neurobiology of learning and memory, 94*(4), 530–537.

Zijlmans, M. A., Korpela, K., Riksen-Walraven, J. M., de Vos, W. M., & de Weerth, C. (2015). Maternal prenatal stress is associated with the infant intestinal microbiota. *Psychoneuroendocrinology, 53*, 233–245.

Zmyj, N., Jank, J., Schütz-Bosbach, S., & Daum, M. M. (2011). Detection of visual-tactile contingency in the first year after birth. *Cognition, 120*(1), 82–89.

Zubieta, J. K., Heitzeg, M. M., Smith, Y. R., Bueller, J. A., Xu, K., Xu, Y., & Goldman, D. (2003). COMT val158met genotype affects μ-opioid neurotransmitter responses to a pain stressor. *Science, 299*(5610), 1240–1243.

Zunszain, P. A., Anacker, C., Cattaneo, A., Choudhury, S., Musaelyan, K., Myint, A. M., & Pariante, C. M. (2012). Interleukin-1 β: A new regulator of the kynurenine pathway affecting human hippocampal neurogenesis. *Neuropsychopharmacology, 37*(4), 939–949.

Zwicker, J. G., Grunau, R. E., Adams, E., Chau, V., Brant, R., Poskitt, K. J., & Miller, S. P. (2013). Score for neonatal acute Physiology–II and neonatal pain predict corticospinal tract development in premature newborns. *Pediatric Neurology, 48*(2), 123–129.

Index

acetylation 90, 91, 94
adrenocorticotropic hormone (ACTH) 57,
    62, 67, 93, 142
alarmins 106
allostasis 65–66
allostatic load 65–67, 106, 152
allostatic load, mitochondrial 106–107
Als, H. 120
ambiguous nucleus 48–52
amygdala 40, 48, 65–66, 103–104,
    133–134, 143
animal models: dog's gesture of licking 28;
    frog heart experiment 33–34; germ-free
    mice and rats 104; kangaroo care 120;
    memory facilitation 69; rat model 56;
    reptiles 48–49; reptiles and mammals
    48–49; sea snail, *Aplysia californica*
    (Kandel) 36–37; *see also* rat mothers
anterior fontanelle 18
anti-inflammatory cytokines 101, 114
arginine vasopressin 131
Australopithecus africanus 18
autonomic nervous system 47, 56–57,
    78, 81
axons 35

BDNF expression 98, 104, 133
behavioural epigenetics 92–93, 96–99; of
    prematurity 126–128
behavioural genetics 72–76, 82–87
Belsky, J. 76–80, 83–84, 118–119
biological sensitivity 78–84, 119
bio-rhythmicity 10–11
Boyce, W. T. 78–80, 119
brain 24, 30, 66, 143; activity 32, 46,
    138; architecture 18–19; brainstem
    34, 40, 46–47, 49, 51; brain-to-brain
    tuning process 139; circuits 35, 132;
    co-regulation 139; development 18–19,

29, 35–36, 103; paternal 132–134;
    plasticity 29, 37–38, 134; volume 18,
    22, 103, 127
Brazelton, B. 120

caregiving: behavior 68, 131–132;
    environment 4, 87, 97; maternal
    130–132; paternal 131–132;
    programming 115–117
carry-over effect 13
catechol-O-methyltransferase (COMT)
    85–86
central nervous system 26, 29–30, 32–35,
    38–41, 46–47, 50, 66, 84, 103, 122–126,
    138, 143, 151
child/children: breastfeeding 62, 104, 123,
    128; dandelion 79; maltreatment 141;
    orchid 79–80; psychosocial stress 102;
    S-carrier 84; sense of agency 15–16;
    temperament and sensitivity 82; vagal
    tone 47–50, 52–54, 137
Chomsky, N. 27
chromatin 90–92
Cicchetti, D. 62, 142–143
complex system 6, 28, 35, 39
contingency 11, 14–16
co-regulation process 13–14, 60,
    135–139
corticotropin-releasing hormone (CRH)
    56–57, 93, 112, 128, 142
cortisol 34, 55–70, 85, 95, 112–114, 116,
    119, 126, 136–138, 145, 147
COVID-19 pandemic 156–158
CpG sites 91, 99, 113, 115, 126–127
C-reactive protein (CRP) 101–102, 144
C-tactile fibers (CT fibers) 25
curved U-shaped relationship 79
cytokines 100–102, 106, 112–114, 142
cytosine 90–91

Darwin, C. 28–29, 46, 88–89
deacetylation 91
dendrites 35, 40
depression 62, 65, 102, 112–116, 142–144
developmental care 121–125, 128–129
developmental origins of health and disease (DOHaD) 111–112
developmental psychology 29–31, 72–76, 86, 152
diathesis-stress model 75–76, 78–80, 82–83, 152
DNA: identification of 71–72; methylation 62, 90–92, 98, 125–127, 145–147; mitochondrial 105–107
dopamine D4 receptor (DRD4) 84–85
dopaminergic system 84–85, 119, 133
dorsal nucleus 48, 52
dyadic regulation process 14–16
dynamic system: non-linear 8, 10, 24, 55, 87, 89, 100; recursiveness 6–7; self-regulation 6; unpredictability 7–8

early life experiences 61, 71, 115
electroencephalographic (EEG) 21, 138–139
Ellis, B. J. 78–80, 119, 154–155
emotional regulation 13–14, 59–60, 82, 85, 105, 127, 136, 143
epigenetic process 62, 90–99, 112–115, 119, 121, 124–126, 128–129, 151, 157
evolutionary adaptation 76–78
evolutionary psychology 28
eye-tracking technology 21

Family Nurture Intervention 123–124
father-child dyad 138–139
Feldman, R. 5, 53, 116, 133, 135–136
fetal/fetus 19–20, 112–119, 134; growth/ development 97, 111; heart rate 20; movements 19–20, 117
FKBP5 gene 97, 144, 148
Freud, S. 26, 32

Galvani, L. 33
gamma-aminobutyric acid (GABA) 34
Gene × Environment (G × E) studies 82–86
general adaptation syndrome 56
genetics and psychology 72–73
genetic variations 75, 82–86, 95–96, 106, 126
Genome Project 89

genome-wide association studies (GWAS) 75
glucocorticoid receptors (GR) 58, 62–64, 69, 93, 113–115, 126, 142, 144, 147
glucocorticoid resistance 113–114, 142
glutamate 34, 39
Gould, S. J. 28
Grossman, P. 51–52
Gunnar, M. 67

habituation 36–37, 53, 59, 64, 113, 137
heart rate 20, 36, 47, 57, 125, 142
heart rhythm 20, 47, 50–52
Hebb, D.: axiom 26, 38; postulate 11; principle 35, 37–38
heritability 74, 86
hippocampus 38, 40, 58, 63–65, 68, 94, 97, 99, 104, 126, 133, 143
histone tail 90, 91
Holocaust survivors 141, 147–148
homeostasis 46, 55, 57, 65
hormones 131
5-HTTLPR (serotonin-transporter-linked promoter region) 83–84, 96, 115, 126–127
human brain *see* brain
11 βhydroxysteroid dehydrogenase (11 βHSD-2) 112–113
hypothalamic-pituitary-adrenal axis (HPA axis) 57–58, 100–101; activation of 60–62; in memory and learning 68–70; in parent-child interaction 136–137; programming of 62–64; regulation 58–59; and serotonergic system 126
hypothalamus 56–58, 65, 93, 131–132, 134

IL-6 101–102, 106, 114
Insel, T. 45
interbeat heart rate 47
inter-neuronal communication 33

Kandel, E. R. 36–39, 150
Karmiloff-Smith, A. 30–31
kinases 39

Lamarck, J.-B. d. 88–90
language development 31
LG-ABN behaviour 62–64, 68, 92–93
life experiences 30, 35, 61, 71, 95, 97, 115
like-me-ness 21
living systems, organization of 8–9
Lømo, T. 38–39

long-term memory 39–40
long-term potentiation (LTP) 38–40, 69

mammals 47–49, 51–52, 102–104,
    131–132, 134
MAOA gene 83
Martha Welch's group 123
maternal: behaviour 62, 64, 68, 84, 92–95,
    98, 130–134, 136; caregiving 130–132;
    cortisol 112–113; depression 112–115;
    malnutrition 111–112; maternal-fetal 20,
    113, 116–118; oxytocin 136; sensitivity
    84–85, 102, 123, 127, 134; separation
    61, 67–68, 104, 122; stress 20, 62,
    104, 112–115, 117; touch 24–25; voice
    19–20, 124–125
McEwen, B. S. 64–65
Meltzoff, A. 20–21
microbiome 102–105, 114
microRNA 92
mineralocorticoid receptors (MR) 58, 69
mineralocorticoids 58
mirror neurons 21–22, 32
mirror system activity 21–22
mitochondria 105–107
mitokine 106
molecular genetics 73–75
mother-child dyad 4, 6, 8–10, 136–139
mu rhythm 21–22
mutuality 14
mutual regulation model (Tronick) 12,
    14, 60

near-infrared spectroscopy (NIRS) 125
neonatal imitation 20–21
neonatal intensive care units (NICUs)
    120–129
neonatal pain management 121–123, 128
neural learning 37–38
neuro-constructivist approach 30–31
neuroendocrine: cascade 57–58, 65;
    response 56, 65, 67, 114–115, 122, 128,
    145; system 60–61, 126, 137, 143
neuron 18–19, 21–24, 32–33, 35–37, 51–52
neurophysiological response 25, 31, 57
neuroplasticity 18–19, 37–38, 115–116,
    125–126
neurotransmitter 33–35, 37, 39–40, 126
newborn 10, 47, 51, 66, 114–120; handling
    61; protection 122–124
Newborn Individualized Developmental
    Care and Assessment Program
    (NIDCAP) 120

NMDA receptor 39–40
noncoding RNA 92
non-linear dynamical systems 7–10, 24, 26,
    35, 40, 55, 59, 87, 89, 100
NR3C1 gene 93–95, 97–98, 113, 115, 126,
    128, 144, 147

open space 14
oxytocin 34–35, 63–64, 98, 115–117, 119,
    131, 134–136
oxytocinergic system 63–64, 134
oxytocin receptor (OXTR) 98, 115

parasympathetic system 47, 49, 54, 57, 137,
    151
parental/parenting 67–68, 77, 82, 85, 87,
    93, 123, 125, 127, 133, 135; buffering
    effect 67–68; sensitivity 68, 116
parent-child system 5–6, 11; coordination
    process 15; dyad 5–6, 8–9, 68, 135,
    139; HPA axis 136–137; interactions 4,
    10–11, 13–16, 23, 135–136; matching
    and mismatching 13–14; meaning
    making 15–16; physical and tactile
    exchanges 24; relationship 15, 17,
    53–54
paternal: behaviour 131–132, 134–135;
    brain 132–134; caregiving 131–132;
    deprivation 133
peptide-type neurotransmitters 34
physical contact 22, 24, 64, 122–124, 126,
    128, 135
plasticity 29, 37–38, 76–78, 80–83, 112,
    119, 132–134, 151–153
polymorphism 75, 83–86, 96, 115
polyvagal theory 45–47, 49–52
Porges, S. W. 45–53
posttraumatic stress disorder (PTSD) 97,
    101, 143, 145, 147
prefrontal cortex 58, 65–66, 104, 133, 139,
    143
pregnancy: blood pressure during 118;
    depression in 113–115; diet during
    94; Hongerwinter effect 97, 146–147;
    hormones 116–119, 131–134; ice storms
    hit during 141, 145–146; maternal
    stress 112–113; September 11 attacks
    144–145; stress response 103–105;
    trimester of 20, 113–114, 118, 136,
    145–146, 157
premature babies 120–122, 126
prematurity, behavioral epigenetics of
    126–128

prenatal maternal stress 112–114
principle of divergence 35
pro-inflammatory cytokines 101–102, 106, 112–114
Project Ice Storm 146
psychosocial stress 60, 65, 85, 100–102, 107–108, 118

quantitative genetics 73–74, 86
quantitative trait loci (QTL) 75

Ramón y Cajal, S. 31–32
rat mothers 92–93; high-quality care (HQC) 93–94; low-quality care (LQC) 93–94
reactive oxygen species 106–107
reciprocity 14
resilience 58–60
respiratory disease 78
respiratory sinus arrhythmia (RSA) 49–53, 124, 137
reunion episode 12–13, 84
rhythmicity 10–11

salivary cortisol 59, 61, 69–70, 85, 103–104, 113, 136–138, 145
Sander, L. 3–5, 8–9, 11, 14, 17, 56, 150
Selye, H. 56, 64–66
sensitization 37, 132
serotonergic system 83, 96, 114, 119, 126, 135
serotonin transporter gene (SLC6A4) 95–96, 98, 115, 126–128, 144, 157
short-term memory 39–40
single nucleotide polymorphisms (SNPs) 75
skin-to-skin contact 22, 120, 123, 128, 135
sleep-wake rhythm 5, 11, 122
social agent 15
socio-cognitive skills 21, 23, 31, 126
specificity 9
Spitz R, A. 4

still-face procedure 12–13, 24, 51, 53–54, 60, 68, 70, 84–86, 127–128, 136–137, 139, 151
stress 56; defined 56; diathesis-stress model 75–76, 78–80, 82–83, 152; evolutionary adaptation 76–78; hormone 34, 113, 116–119, 131–134; LG-ABN behaviour 62–64, 68, 92–93; maternal 20, 62, 104, 112–115, 117; neuroendocrine cascade 57–58; psychobiological mechanisms 78–79; psychosocial 60, 65, 85, 100–102, 107–108, 118; resilience 58–60; response 56–58, 67, 70, 85, 103, 126–128, 142; vulnerability 75–76; *see also* hypothalamic-pituitary-adrenal axis (HPA axis)
susceptibility 77–82, 84, 118–119
sympathetic nervous system 46, 49, 57, 65, 100, 137
synchrony 14, 52–53, 82, 117, 133, 135–139
system, defined 5–6

Taung (South Africa) 17–19
telomere 106–108, 128, 144
trauma/traumatic experiences 83, 86–87, 114, 141–148, 157
Trier Social Stress Test (TSST) 68, 142, 144
Tronick, E. 3, 11–12, 14, 59–60
trophic cascade 6
tumour necrosis factor alpha (TNF-a) 101–103, 113–114
twin registries 74

von Economo neurons (VEN) 22–25

Waddington, C. H. 89–90
Weiss, P. 8–11

For Product Safety Concerns and Information please contact our EU
representative GPSR@taylorandfrancis.com
Taylor & Francis Verlag GmbH, Kaufingerstraße 24, 80331 München, Germany

*9 7 8 1 0 3 2 7 6 6 1 5 7 *